SECRETS OF
123 OLD-TIME SCIENCE
TRICKS & EXPERIMENTS

No. 1261
$12.95

SECRETS OF 123 OLD-TIME SCIENCE TRICKS & EXPERIMENTS

by Edi Lanners

TAB BOOKS Inc.
BLUE RIDGE SUMMIT, PA. 17214

FIRST NORTH AMERICAN EDITION

FIRST PRINTING—DECEMBER 1980
SECOND PRINTING—MAY 1981

Library of Congress Cataloging in Publication Data

Lanners, Edi.
    Secrets of 123 old-time science tricks & experiments.

    Includes index.
    1. Science—Experiments. 2. Scientific recreations.
I. Title.
Q164.L26 507'.8 80-21080
ISBN 0-8306-9675-X
ISBN 0-8306-1261-0 (pbk.)

Original by Verlag C.J. Bucher, Lucerne and Frankfurt, 1976. First
English Language Edition by Paddington Press (U.K.) Ltd., 1978.
Translation of Kolumbus-Eier, based on a larger collection entitled
Columbus Egg, first issued by the editors of Der Gute Kamerad.

CONTENTS

INTRODUCTION

MY dear friend,

A while ago I mailed you two old volumes of indoor games entitled *Columbus' Egg*, together with my suggestions on how they might be adapted for possible republication. I asked for your opinion on whether a book of this kind could hope to succeed, bearing in mind the veritable flood of contemporary works.

I was quite unprepared for the enthusiasm of your reply. I had looked forward to your opinion on the book as a text, but never expected your wife and children actually to put the tricks to the test. The fact that you tried many of them out together is fully in the spirit of the book – it is meant to be used as well as read. I have a shrewd suspicion, however, that you hoodwinked your family into large-scale experimentation simply to find out whether the delightful pictures bore any resemblance to the detailed instructions. I bet you felt that you would be sending me back a clutch of rotten eggs!

I must now confess that we put ourselves through the same sort of fun and games, and stood Columbus' eggs on end by the dozen. We regularly misappropriated household utensils, upset the cooking, interrupted mealtimes, turned social occasions into demonstrations of our skills, and our youngest son even tempted our neighbor, a dignified old gentleman, into trying his hand at childish experiments with knives and corks.

But now for your "whys" and "wherefores." The publishers originally asked me to collaborate on a book of magic. Publications on that theme usually address themselves to individuals, and are hence, I feel, unlikely to encourage group participation. The sorcerer's apprentice, after all, relies on the ignorance of his audience. If he serves up tricks that are simple and familiar, he will undoubtedly give the children pleasure and hold their interest for some time, but if he starts to reveal the secrets of the more intricate conjuring acts he is bound to strip the latter of much of their fascination. In these cases what matters is a perfect performance, not the explanation. Constant practice and endless patience are needed to hold the attention of a passive audience; only with these may one hope to turn an illusionary trick into a spellbinding spectacle. It is difficult, however, to convey all this on the printed page, so that a book addressed only to the individual would-be magician is likely to prove a source of discouragement. We decided, therefore, to elicit the participation of the entire family.

It was this consideration that led me to the old models. I had discovered the two volumes of *Columbus' Egg* some time before in a secondhand bookstore. The owner had given them to me when I purchased a book on butterflies, voicing the opinion that it was unlikely for anyone to evince the slightest interest in such childish games now that the toy industry is swamping the market, year after year, with a stream of new and ever more marvelous playthings.

The year of publication, 1890, was hidden in the vignette of the first volume. We may take it that the book was a compilation of articles published separately in various journals; I discovered some of the old wood engravings in German and French almanacs, together with different, often much lengthier, instructions. Other pictures, illustrating or explaining scientific experiments, have turned up in old textbooks. For two generations they apparently served our great-grandfathers and great-great-grandfathers as models for games or as educational aids at home and at school, later to be ousted by the rise of photography and more abstract illustrations.

You expressed your surprise at the craftsmanlike and detailed execution of the illustrations which, judged by present-day standards, bear no economic relation to the text. In other words, why was so much time and love spent on depicting, say, a carafe and straw modeled in light and shadow so carefully and realistically, if all that the picture was intended to illustrate was the strength of the straw?

Nowadays we have come to rely largely on photographic illustrations which, as a convenient and efficient method of reproduction, no longer presuppose the sensitive involvement of the artist with the object to be depicted. Moreover, we have also become used to a kind of shorthand and consider repetition to be superfluous. In daily life, we presuppose as generally known many things that may long since have been forgotten. The old illustrations and texts, by contrast, keep harking back to the antecedents of a given theme and do not hesitate to repeat details that are seemingly of secondary importance.

Thus if we compare an old introduction to physics with a new one, we shall be struck by how humorless and puny our abstract signs and symbols seem beside the amusing and vivid old illustrations. A force whose direction and magnitude is nowadays represented by an arrow would in earlier times have been depicted much more tellingly by a flexed and powerful arm or by a team of plodding draft horses. A child can grasp the elemental force of the wind when it is personified by an angel blowing as hard as he can, or by an enraged god of storms using his hurricane breath to drive a huge sailboat through the ocean waves. In short, science used to be not only easy to understand, but also amusing and stimulating. Indeed, play was generally thought to pave the path to knowledge.

The nineteenth century, to which we owe our scientific games, still allowed the gods to make themselves felt, and explored and explained creation with their help and "collaboration." However, we should be quite wrong if, in nostalgic retrospect, we imagined that the age was a divinely protected era of innocent recreation.

By that time man's questing intellect had already linked the wheel to steam power, and the world was being transformed, as never before, with incredible speed: a constant stream of new and ever-better machines drove a wedge between man and his familiar tools. Industries sprang up that required strange new procedures with levers and wheels; cities grew in size, and were brought closer together by ever faster traffic. . . .

In 1830, the diligent draft horse celebrated its last victory in a race with a locomotive along the Baltimore-Ohio track. Some ten years later, the British railroad system could boast a length of more than 2,000 miles. In 1859, oil appeared as a new source of energy alongside steam power, and as early as 1881 the first electric streetcar was conveying passengers in the vicinity of Berlin.

Small wonder, then, that people everywhere were making great efforts to understand the new techniques and to wonder how best to exploit the latest advances. The daily press published innumerable proud and detailed accounts by scientists and engineers of their inventions and discoveries, and the second half of the nineteenth century embarked upon a cramming course in physics based on simple examples, realistically and popularly presented. It is against this background that we may begin to understand the approach used in this little book, an approach that is no longer familiar to the modern reader.

The present compilation invites the reader to explore "strange continents" inside the family home and awakens the Columbus slumbering within us all. This object lesson in the form of family games using ordinary household articles can give rise to more pleasure and understanding even today than many a heavy tome eking out its prestigious existence unread on our bookshelves.

Apart from a few exercises in dexterity, I have only included simple, elementary games illustrating equilibrium and the center of gravity, inertia, centrifugal effects, the behavior of liquids and gases, simple optical and acoustic phenomena, electrical forces and other easily duplicated experiments.

I agree with you when you say that man gains his best insights when he is in contact with everyday objects. This view is corroborated not only by the celebrated apple which fell on Newton's head while he was having an afternoon nap, and to which he owed his discovery of the laws of universal gravitation, but also by the ingenious Eiffel Tower, which could hardly have been built for the Paris Exhibition of 1889 without some understanding of the structural properties of such apparently simple objects as straws.

How marvelous it would be if the most fertile brains of this century, too, had been able to provide similarly splendid illustrations of what goes on around us! Far too often we stand undiscerning and powerless before the manifold problems we face, and wish devoutly that we were granted greater understanding. Accustomed to life with the mysterious atom, fearful of its applications in peace and war, we lack popularizers and interpreters of man's latest achievements and discoveries. Without these "go-betweens," we resemble the blind who trustingly follow or fervently condemn without stopping to consider that they may be lacking in basic insight.

How many such games could we not do with today, games that reveal not only the mysteries of physics but also of man himself?

You ask me about the method of reproduction used in this book. As I said earlier, all the old-fashioned and charming pictures are wood engravings, or xylographs, which in the middle of the nineteenth century used to make up the greater part of all illustrations and were widely used, for instance, in illustrated journals. Meyer's *Universal Encyclopedia* at the turn of the century bears the following entry:

> Xylography is the art of cutting out of wood blocks, designs that have been previously sketched in with pen, pencil or brush, or reproduced by mechanical means, for instance, by photography, in such a way that the designs are suitable for printing. The procedure is as follows: after the wood block, with a standard thickness of $\frac{5}{8}$ inch, has been planed level and smooth on one side, it receives a thin coating of chalk. The design is applied to this coating, but in reverse, as a mirror image of the future print, or else it is projected onto the block by photographic means. The block is then handed over to the wood engraver, who cuts out all the areas the draftsman has left untouched, with a graver or solid burin, so that when he has completed his work the drawing alone will stand out. The block is now spread with printer's ink and, when applied to paper, will produce a mirror image of the original. Apart from technical competence, which requires long training and practice, the xylographer must also have a considerable degree of artistic sensitivity and be himself an accomplished draftsman, especially when dealing with designs that do not consist of lines but, like most of the blocks used by illustrated papers, have been painted or brushed on.

Most of our illustrations came from French studios. Amongst the signatures I discovered that of the Parisian Henri Thiriat, one of the most important wood engravers in France.

You maintain that most modern children do not learn games from their parents and that they participate in competitive games as passive spectators. I should not put it quite so bluntly. But I agree that, if they do not take part in spontaneous games, children become stunted in their intellectual development, a statement that probably applies to adults in equal measure. Even without our intervention, the growing child makes tentative attempts at unfettered creation and, by painting, building and modeling, erects bridges to his or her environment. We adults value games and play far too cheaply; in addition to a few evenings in the bowling alley or at the card table, we are at best involved in just one kind of sport, and the lazier and less socially inclined amongst us are quite satisfied to sit back and do nothing but watch our televisions. Psychology teaches us that game playing covers a much wider spectrum than we generally think. It is a self-rewarding activity, one that has no ulterior motives. In their rhythmical comings and goings, in their encounters with objects and society, their involvement and activity, players see themselves as being in immediate and significant contact with the mainstream of life. Roles, imaginary situations and illusions are acted out from the second year of life onward, the young child still failing to grasp the make-believe character of his or her games. Playing comes into its own in the preschool and early school period. However, creative play continues to be the basis of our working life, leading beyond youth to adult games and occasionally taking on the character of serious intellectual, artistic and industrial activity.

In conclusion, I may perhaps be allowed a few words about the title; the legendary egg of Columbus was probably foisted like a veritable cuckoo's egg, even during his lifetime, onto the great Genovese, who discovered America under the Spanish flag. As is well known, we refer by a "Columbus's egg," to an astonishing solution or completion of an apparently insoluble problem or task; to a creative idea that overcomes all obstacles. Our titular hero, Christopher Columbus (1451–1506), being convinced of the spherical shape of the earth, intended, by following the curvature of the earth, to reach the east coast of Asia while sailing toward the West, although that part of the world had never before been reached from that direction, and although most of his contemporaries still held the ancient belief that the earth was flat.

After his voyage, which despite many setbacks he completed successfully, his achievements were challenged and his detractors scoffed that others could undoubtedly have made the same discovery.

It is at this point that the legend comes in. It tells us that Columbus put the following question to his denigrators: "Can any of you stand an egg on its head?" When they had all responded with embarrassed silence, Columbus is said to have stood the egg on the table on its crushed end, mortifying and astounding the braggarts with his simple solution.

In reality, this trick was devised, not by Columbus, but by one of his compatriots, the brilliant architect, Filippo Brunelleschi (1377–1446), who, as Vasari reports in his *Lives of the most eminent Painters, Sculptors and Architects* (1550), offered it to his opponents when they declared that his proposed cupola for the cathedral in Florence was a castle in the air. This revision, however, in no way detracts from a legend whose effect does not depend on whether it was devised by Brunelleschi or Columbus. The crucial point is the challenge it poses: not to consider everyday tasks as a chore, but to meet them in a creative spirit, in play no less than in real life.

Good-bye for now, my dear friend,

Yours affectionately,

Edi Lanners

P.S. Does work come before play, or does play come first?

COLUMBUS'S UP-DATED EGG

LEGEND has it that, at a dinner given by Cardinal Mendoza, those who had previously scoffed at Columbus were dumbfounded and humiliated when he successfully stood an egg upright by tapping its end, a feat of both physical and mental agility. Students of the human psyche will conclude that Columbus must have employed a hard-boiled egg for it seems unlikely that the alleged effect on the illustrious company would have been quite so overwhelming and unequivocal had a raw and fragile egg, woefully deprived of its shape, besmirched the clerical table. But let us leave Columbus' solution, admittedly impressive but comparatively banal, to penetrate further into the mysteries of equilibrium by asking: "How can a raw egg be stood on its head?"

The reader will smile condescendingly, for he or she commands quite a few more or less elegant drolleries and subterfuges. Thus, all you need do is shake the egg vigorously until the white and the yolk have been thoroughly mixed. If the egg, now divested of its divinely ordained inner order, is held upright for a time between the fingers, the heavier yolk will sink to the bottom, lending the lower part so much weight that, if the egg is set down on a table and balanced with some patience, it will remain in an upright position.

Not so simple, but by no means difficult, is the experiment depicted in our illustration.

The top of a cork is hollowed out so that the blunt end of an egg fits neatly into it. If the cork is now placed over the egg and weighed down on either side with a fork, the center of gravity of the whole system is shifted downward. After several attempts, and slight changes in the position of the cork and the egg, stable equilibrium is attained, with the result that, as the illustration shows, the egg and its burden can be balanced gracefully on the rim of a bottle – to the great astonishment of the assembled company.

THE HOVERING LADLE

HOVERING is a most mysterious business. Does a bird hover? The commonly held view is that it does, but strictly speaking it does not. It "rests" on the air, for if it were in a vacuum the miraculous powers of its wings would desert it, and it would plummet to the ground. Hence we must not take it amiss if the ladle and the knife, which are here supposed to suggest the idea of "hovering," need a point of support. Let us give it to them and stick the point of a half-opened pocketknife, held in a horizontal or near-horizontal position, into the edge of the table. As is shown in the illustration, it must be imbedded just enough to support the ladle. The position adopted by the two "performers" is most intriguing, especially when they are set swinging very gently.

The spectators will invariably be amazed. We, who are privy to the secret, will give a superior smile, for we know that the spectators' astonishment will increase even further when we pour a little sand into the ladle, or weigh it down in some other way.

"Will it rise or sink?" we ask our audience.

The unanimous reply is naturally that it will sink, as any object that is weighed down tends to do.

Our two "performers," the knife and the ladle, however, seem determined to play a joke; the first rises in the air, and the ladle goes with it, and the more we weigh the latter down the more upright a position will the former adopt, until it reaches a certain point, ends the performance and falls off the edge of the table with a clatter.

THE ACROBATIC PENCIL

TAKE a pencil, not too short, but otherwise of any shape, and push in

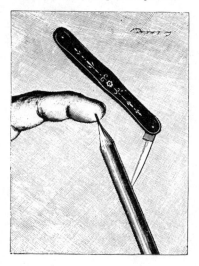

the point of a penknife (or any other knife) not far from its tip. Gradually bring the handle toward the point of the pencil, until the latter, resting on the index finger, remains in balance. This happens just as soon as the center of gravity is beneath the point of support, that is, underneath our finger. The slope of the pencil depends on the angle between the handle and the blade of the knife; the pencil will be vertical when the center of gravity of the knife is in line with the pencil. Those having exceptionally sensitive fingers should ensure that the pencil is not too sharp, for, weighed down by the knife, its point will press quite deeply into the finger tip; you may, of course, introduce a small piece of leather or cloth in order to minimize that danger.

Incidentally, the experiment can be performed just as successfully with a match instead of a pencil; indeed it gains in appeal thanks to its trifling size. Nor is that all: we can balance another match on the end of the first, but in this case it is best to use a smaller knife. With a little skill, the game can be further extended, and a third match and knife placed on the first two, thus producing a veritable tower which leaves nothing to be desired in the boldness of its execution.

Care must, however, be taken, for, as the structure grows, the blade may snap shut by itself, thus causing a disagreeable as well as unwelcome injury to the finger. It is better, and safer, if the "tower" is erected on the cork of a stoppered bottle.

The Plate Carousel

IF we ask the good lady of the house for a plate, informing her that we shall be trying to balance it on a needle, it is ten to one that mindful of her crockery, she will refuse our request. But if she has had occasion to notice that we sometimes perform a well-nigh impossible trick without causing the slightest damage to even the most fragile objects, she may yet

be persuaded to lend us the precious plate, and, with some trepidation, the dangerous needle as well. We shall need in addition a stoppered bottle, two corks and four matching forks. These objects, too, are easily come by, and as soon as we have them we can proceed to the construction of our needle-borne plate merry-go-round.

To begin with, we obtain four half-cylinders by cutting each cork in two lengthwise. The cuts must be made with a very sharp knife, otherwise the cut surfaces will be too rough. If this is unavoidable, then we must carefully smooth the cuts with a file. We now push the four forks into the ends of the cut surfaces of the corks, taking care that they are not completely perpendicular to the latter; the illustration shows the approximate angle very clearly. It goes without saying that the inclination of all four forks must be roughly the same. If our preparations have progressed satisfactorily thus far, then we may load the plate with the four forks, placed at the four points of the compass around the edge, and finally place the plate on the blunt end of the needle, whose point we have pushed vertically into the cork of the stoppered bottle. After several adjustments, we shall discover the point at which the whole system is in equilibrium; indeed, with some care – for instance by blowing or gently tapping – we can even set the plate in motion and cause it to turn like a merry-go-round, slowly at first and then faster and faster until the good lady of the house is proved right after all and she has to lament the loss of a plate, consoling herself, perhaps, with the thought that she has sacrificed it to further the noble art of equilibration.

FLOATING KITCHEN UTENSILS

THIS simple experiment, too, can demonstrate that it is possible, with equipment readily found in most households, to perform the most astonishing balancing feats, if only we are inventive enough and give free rein to the imagination.

First, we take a decanter (or a similar vessel with a high neck), which we fill with water, and also a plate, a skimming ladle and a soup ladle, each of whose handles must have a hook. Thus equipped we return to our experimenting or demonstration table, attach the skimming ladle by its hook to the edge of the plate and wedge it there with a slice of cork. Next we pick up the two connected objects in one hand and place the edge of the plate – bottom upward – on the rim of the decanter. With the other hand we hang the hook of the soup ladle over the edge of the skimming ladle and gradually, by making careful adjustments, discover the position at which the whole system is in stable equilibrium.

The audience will not be slow to applaud, and we may finally call in the lady of the house so that we can demonstrate, to her astonishment, that her kitchen utensils can serve for purposes quite other than the production of the occasional sirloin roast of beef with cream sauce.

Playing with Matches

The tricks that follow, for which we need a box of matches, are again both very simple and most effective. We begin by splitting a match very carefully at its non-striking end; the end of another is then cut into a wedge and positioned in the slit of the first, so that the two form an acute angle. Next, we place the two interlocking matches on the table and lean them against a third, so that all three stand up as shown in the lower part of the illustration on page 18. All the preparations for our little entertainment are now complete, and we ask if anyone thinks he or she can lift up the whole framework with the aid of a fourth match. There will be much shaking of heads, and many vain attempts, since the only possible solution is

the one shown in the upper part of the same illustration: the third match must be placed between the joined tips of the first and second matches and the whole then raised with the fourth match.

Providing our public shows just a little appreciation, we may treat them to an amusing encore. A match is placed on the table, and an even number of others laid across it, each with one end resting on the table and the other pointing upward; the heads must point alternately in opposite directions, as shown in the illustration on page 18. We again ask our audience to use a single match to lift up all the others in the order in which they are arranged on the table. It is unlikely that anybody will consider this task too difficult; those

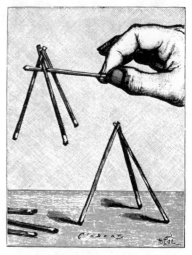

present will try to solve the problem in many and various ways – but in vain! If they do not know the trick the matches will collapse in utter confusion every time they are touched. We, for our part, now place the match we are allowed to use parallel to and on top of the one at the bottom. If we take hold of the lower (first) match, we are able at once to lift up the rest in the required manner, the unsuccessful contenders being inordinately incensed at not having hit upon this extremely simple solution themselves.

Once we have performed this trick, we can add yet another little diversion. We state that with the help of some of our matches we can turn ink into oil. The bystanders will smile incredulously, and perhaps even lay a wager, which needless to say, we win, for we quickly place seven matches on the table, thus:

The gamblers will by now have begun to feel rather uneasy, for no doubt they will have guessed that we are about to effect the transformation by a simple rearrangement of the matches, to wit:

For variety, and particularly in the presence of those who are familiar with this trick, we can think up quite a few other examples, with the proviso that all the pairs must have an equal number of matches, and spell out distinct concepts.

A Tiny Drill

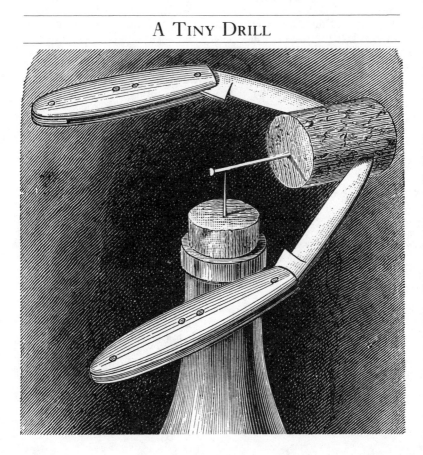

Is it possible to drill a hole through a nail, simply by blowing? No, it is not, we shall be told; you must use a sharp instrument, or better still a lathe or a special drill. Quite right! So let us make use of equilibrium and the law of gravity to manufacture just such an implement. We only need a few objects, to be found close at hand: a needle, two penknives of equal weight, a corked bottle and an extra cork.

We begin by sticking the nail into the spare cork, and the blunt end of the needle into the bottle cork. Next we open both penknives and stick them into the spare cork on either side of the nail, as shown in the illustration. We must now make sure that the center of gravity of the two knives lies exactly at some point in the nail. To that end, we place the latter on the tip of one of our fingers, and ask an assistant to move the knives to and fro until equilibrium is attained. Our preparations have been completed.

We now place the entire structure on the point of the needle, previously inserted in to the bottle cork. Once again we must try to find equilibrium by carefully and lightly touching and moving the structure. Just a modicum of practice is needed for success. Once the nail and its load are in balance, we begin to rotate the two knives, the cork and the nail by blowing at them. We have kept our promise: because the needle is of tempered steel and the nail of softer metal, the former bores gradually into the latter like the finest of drills, producing an ever deeper hole until the rotating nail is finally bored through.

THE DANCING WINEGLASS

WE obtain two empty wine bottles of equal size and cork them in such a manner that the corks protrude to the same extent from the necks of the bottles. The tops of the two corks have previously been cut into wedge shapes. For our experiment we also need two knives with fairly heavy handles, which we now lay across the wedge-shaped corks so that the tips of the knives touch. Where they meet, we now stand a fine glass goblet and pour water or some other liquid into it until the glass is balanced by the weight of the handles. So much for the preparations; now for the actual performance of our experiment: if a small metal sphere, a coin or a button, is attached to a thread and dipped into the liquid, the glass standing on the tips of the knives will sink down, only to readopt its earlier position as soon as the sphere or coin is lifted out of the glass. This process can, of course, be repeated time and again and has a most agreeable effect.

Moreover, we can lend the experiment even greater charm if we invite our host's child to sit at the piano and play us a nice little dance at not too quick a tempo. By dipping and lifting the suspended metal object, we are able with ease to allow the glass to sway in time to the music, thus affording our public a spectacle that is surprising and charming to boot.

THE BALANCING GOBLET

In this demonstration and those that follow, we shall be engaging in experimental physics, and will prove that it is possible, with some ingenuity but without any special preparations or complicated equipment, to demonstrate intriguing physical effects in the simplest of ways, thus not only entertaining our audience but at the same time stimulating their intellects with instructive displays.

Let us begin with an easy but nonetheless elegant experiment, based on the theory of equilibrium. Our task will be to balance a wineglass on the ends of three sticks and to that end we take out three matches. We pick up two of them, cross them and hold them firmly between two fingers of our left hand, making sure that the top match is horizontal. Then we push a third match obliquely under the upper half of the vertical, and over the right side of the horizontal match. In this position, we place them on the table, and shift them slightly until they form an equilateral triangle in the middle. Now the small improvised stand is ready and will support any object of medium weight without giving way or collapsing. Each match has one end on the

table, passes beneath another match, rests on a third and has its free end in the air.

The construction of a similar device of larger dimensions – with rods, rulers or knives – should pose no further problems. If we again place an object of some weight on it, an ashtray, for instance, or – as our illustration shows – a delicate wineglass, it will not be endangered in any way, for the solidity of our slight structure is by no means reduced by by the extra burden, but rather enhanced with the increased pressure.

THE FLOATING CARAFE

The variation of the previous experiment, depicted on the facing page, is no less interesting to perform. The carafe appears to be floating in midair and yet the three knives provide

a relatively firm structure. Our illustration shows clearly how the knives must be placed; their arrangement corresponds exactly to that of the matches described before. We

proceed as follows: a knife is first held in a vertical position, the blade pointing to the right; a second is placed across it horizontally, the blade on top, and the third is inserted between the blades at the top right. The knives are not held by their handles but by the backs of the blades, so that the blades themselves face inward. The blade of the third knife passes under that of the first and over that of the second. When we have made these preparations, we place the handles of the knives on the rims of three wine-glasses, and this structure, too, can now easily accommodate a fragile vase with no risk of breakage.

As if this were not spectacular enough, the experiment (as so many others described in this book) can be combined with another balancing trick, say, with a slight variation of the one described on page 13, in which we balanced an egg on the rim of a bottle. To do this, we stick three forks into the cork on top of our egg, their positions corresponding to those of the three knives. After several trials and slight modifications of the positions of cork and egg, we will reach the required equilibrium. The cork must be slightly hollow at the bottom so that it fits the egg fairly closely. This extremely delicate construction will present a most astonishing spectacle!

We cannot, of course, guarantee that no mishap will befall the egg, if it is raw, or the carafe. We would advise, therefore, that the carrying capacity of the knives be first tested by appropriate pressure of the fingers.

A HOMEMADE PRECISION BALANCE

AFTER all these parlor tricks, we shall now proceed to the construction of a practical object: a precision balance, with which we will be able to determine the weight of small and even minute quantities.

As a support, we select a square bottle, some eight inches in height, made of transparent glass and with a broad neck, in the rim of which we have cut a small groove with a tri-angular file (see *b* in Fig. 3 on page 25) diametrically opposite the tiny dimple or hole *a* made with the point of the file. Our scalebeam is a knitting needle, with half a cork pushed to its center in which two sewing needles are inserted in such a way that their ends fit into the hole *a* and the groove *b*. The pans are cut out of smooth, stiff card-board and hung by silk thread or fine brass wire. To attach the pans to the beam we again use two half corks pushed over the ends of the knitting needle (see Fig. 3) and move them to and fro until the equilibrium of our precision balance is finally established.

As a pointer or indicator we use a stiff wire running vertically through the central cork of the beam, with two strips of paper attached to its lower end to facilitate the reading of the position of the pointer inside the bottle; to the top of the wire we attach a small cork which can be moved up and down at will and serves as a regulator. When the beam has come to rest, we scratch two lines with the file on opposite sides of the bottle; they represent the zero point.

Now, while it is possible to buy an official set of weights for our home-made balance, and with their help to file a scale into the side of the bottle, the following procedure is fairly

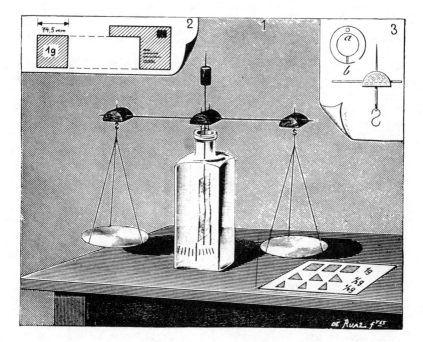

accurate and certainly much cheaper: take a piece of card from a sheet weighing, say, 180 grams per square meter (the usual weight of a post-card), and cut out the requisite number of squares, each with sides measuring 74.5 mm; every square will weigh 1 gram (see Fig. 2). By dividing the squares diagonally, we can produce weights of $\frac{1}{2}$ gram (500 milligrams) and $\frac{1}{4}$ gram (250 milligrams) without any difficulty. For still smaller weights the balance may not be sensitive enough, but just in case we can always make squares with sides measuring 23.5 mm, each of which will weigh $\frac{1}{10}$ gram (100 milligrams). (If desired, coins may be used as troy weights: one U.S. cent weighs 48 grams, that is, exactly $\frac{1}{10}$ ounce.)

THE NERVE-RACKING GAME

OUR illustration shows a balancing game, or, more precisely, puzzle, to which its inventor, doubtless bearing the weak nerves of mankind in mind, gave the appropriate name of "The Nerve-racking Game." The basic equipment is a convex metal lid with a raised edge and a hollow in the middle. A small ball (say, a marble) that fits exactly into the central hollow, is placed on the lid, which is then moved to and fro with the object of rolling the ball into the hollow. This task may be solved in a number of ways, but they all demand a measure of agility. Experienced players allow the ball to climb up the convex surface with a slight initial jerk, at the same time inclining the lid in such a way as to roll the ball up the inclined plane until it reaches the center.

There is, however, a relatively simpler solution, and one that allows the ball to reach its destination more quickly. Let us assume that the ball is very close to the edge: now, instead of trying to take the ball to the hollow, act as if you were trying to bring the hollow to the ball. In other words, move the lid in exactly the same direction as the ball advances. This action, for which the precise amount of force must, of course, be used, is akin to, say, the action of shoveling something out of

the ground. With this ruse, the puzzle can always be solved, no matter where the ball is placed on the lid. Once we have acquired the requisite skill, the trick can be performed even with closed eyes, provided the original position of the ball is known. It is best, generally, not to use both hands but to hold the lid loosely and easily in one hand, and to manipulate it in that fashion.

PIGS IN CLOVER

A PUZZLE similar to the last originated in the ingenious United States. It has the delightful name of "Pigs in Clover" and its solution, too, demands not a little patience. The toy, which anyone capable of cutting paper can easily make, consists of a flat, round cardboard box that may be decorated on the outside in any way desired. Inside, on the circular base, two walls are affixed parallel to the circumference at equal distances from it and each other, each wall having a "gate" exactly opposite the other. At the center of the box is a small, round, covered "house," open to one side, which the inventor has designated the "sty."

The object of the game is, by cautious manipulation of the base, to move the "pigs," represented by tiny balls or marbles of a suitable size, from the outermost ring into the sty, that is, into the covered box.

This puzzle is far from easy to perform and great difficulties may be encountered, particularly by the novice, for even if the first two or three pigs are successfully herded into the sty, the last – the usual number is four – often proves very stubborn, rolling about and engaging in capers that in no way meet with our intentions. And then, just as the last pig is close to home, the slightest of clumsy movements will suffice –

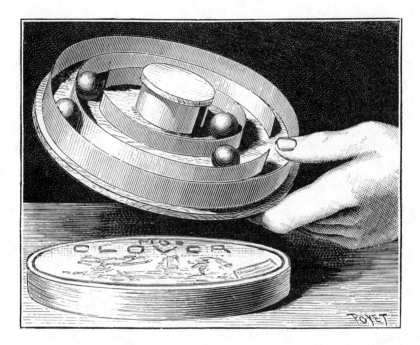

despite our efforts – for one or more of the pigs safely gathered in, to break out of the sty once more. Needless to say it is possible to have any number of pigs, but the larger the herd, the more difficult the task and the requisite measure of patience.

THE ANIMATE CARDBOARD BOX

THIS toy consists of a small oblong box with rounded ends measuring approximately $2'' \times 1'' \times \frac{1}{2}''$, which anyone can make – with a little ingenuity – by following the model depicted on p. 28. If it is painted, or otherwise decorated, the attraction of our little box is greatly increased, but even with an unpretentious appearance it will prove capable of the most remarkable achievements. For inside we place a lead ball whose agility, greatly enhanced by the shape of the box, produces a most singular effect: if the little box is laid on its side at one end of a movable surface, say a household tray, and that end of the tray is raised, the box will scurry to the other end, mysteriously somer-saulting as it moves along, to finish up at the raised edge of the tray in a strange position, often vibrating and shaking like a nervous animal. If the tray is now lifted at that end, the little box will start on its return journey, and if, once it is halfway there, the direction of the tray is changed once again, the box will come to an irresolute stop, before following the new course.

The overall impression is astonishing in the extreme, and this home-made marvel costs next to nothing to produce!

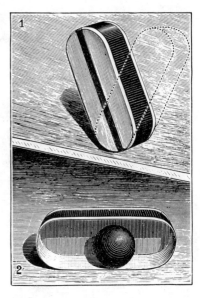

WHERE IS THE CENTER OF GRAVITY?

THE center of gravity of most regular figures is well known; that of others may be easily computed; but for a great many, and particularly for irregular figures, it cannot be stated in advance. Imagine, for instance, cutting Texas very carefully along its borders out of a map of the United States. Where is the exact center of gravity of the resulting piece of paper? Which town can boast that it is the center of gravity of Texas? This question is easily answered by experiment, and shows – but let us not anticipate, and rather describe how the sought-after center of gravity can be found.

We had best start with a simpler case. We moisten the tip of a pencil and draw a parallelogram with two diagonals on a piece of paper. Next

we wet the entire surface of the parallelogram (that surface only, not the whole piece of paper) which is easily done by dripping water on to it very carefully. We now float the whole sheet of paper in a bowl (see the illustration) and gently touch any part of the moist parallelogram with a pin. The floating paper will immediately move until the inter-section of the diagonals comes to lie under the pin, and if we move the pin the paper will respond. The reason is that the tip of the pin is pulled in different directions by the adhesive force of the water on the paper, and equilibrium is only reached when the water is distri-buted evenly around the pin, that is, around the center of gravity – in the case of our parallelogram, the

intersection of the two diagonals. If now, to answer our original question, we draw the outline of Texas or of some other state or country on the piece of paper instead of the parallelogram, and float the surface of the drawing on water in the manner we have indicated, we shall find its center of gravity with no difficulty at all.

Seeing Is Believing

Here is a most diverting experiment, one that has caused great hilarity wherever we have demonstrated it, especially in a company of somewhat corpulent ladies and gentlemen of the older generation! If you wish to try it you had best arm yourself with all the willpower at your command, take up a position a good eighteen inches in front of a wall and support your body by leaning your head against the wall and your arms on a chair. Then try to lift the chair off the floor with both hands, and stand up straight. You will invariably fail to do so, hard though you may try, since the center of gravity of your body in the position described (with the chair off the floor) has traveled up toward your

shoulders, and the major part of your weight is now supported by your head, as the gentle reader will immediately and strikingly discover during a performance of this experiment. If, however, you refuse to accept your helplessness, and insist that – in spite of the laws of gravity – you can return to an upright position, then, even if you are allowed to dispense with the chair and make powerful jerking movements, you will be proved utterly wrong – at best you will come away with a fine bump on your head.

PEELING AN ORANGE

FOR the following trick we need an orange and a sharp, pointed knife. The four diagrams below illustrate the different phases of our procedure. First we make a few simple cuts in the orange peel, as shown on Fig. 1; then we make a series of incisions parallel to the first (Fig. 2) so that we end up with several narrow strips. The next step (Fig. 3) is to join the side strips to the central double strips at the top as well as at the bottom. Finally (Fig. 4), we cut the entire orange very

delicately horizontally and through the middle, taking care not to damage any of the strips, so that we are left with two equal halves.

Needless to say, the cuts shown on Figs. 1–3 must also be made on the other side of the orange, which is not illustrated. If everything is done correctly and the separation of the two halves completed properly, then the sections will automatically come apart displaying the most remarkable entanglements as shown in the enlarged illustration on the right.

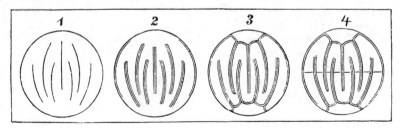

31

The Bewitched Fidibus

"WHAT will you bet me," a friend once asked me, "that you will not immediately drop this fidibus, or paper spill, when I light it at the top?" Curiously I examined the paper my friend held out to me. He had cut out a piece about one foot long and an inch wide, had carefully folded it across in the middle and now held the two ends together between his thumb and his index finger, so that the spill pointed upward.

"Very well; I'll bet you a bottle of wine," I said a little cautiously, aware that without the bet he would not surrender the secret.

My friend now handed me the ends of the strip, struck a match, and lit the spill at the crease; hardly had he done so than the end was burnt through, the paper strip divided at the top, and the two burning pieces of paper spiralled down with lightning speed onto my hand, so that I dropped them in alarm.

The bottle of wine was lost, but what exactly had happened?

While the strip of paper was still uncreased, my friend had pulled it tightly over a sharp edge, thus giving it some tension and causing it to roll up into a spiral. Next he had placed the ends of the strip together in the direction opposite to that in which it tended to curl up, creasing it in the middle so that the spill described above was formed. As soon as the crease had been burnt through, the two ends shot down separately by virtue of their elasticity, and forced me to drop the spill.

I may have lost the bottle of wine, but I more than made good the loss by repeating the wager with quite a few other friends!

THE COMPLIANT COINS

FOR this experiment, which requires no small measure of agility, we need just a dozen small coins which we carefully place in a pile on a plate.

We now ask our audience if anyone can place the coins on the table in the exact order in which we have arranged them, without touching them by hand.

The answer is a general shaking of heads.

"You're going to use another of your tricks, a Columbus's egg or whatever you call it," a skeptical voice may be heard.

"Indeed, the success of the experiment does depend on a slight ruse, but, as you know, I do not engage in sleight of hand, but try to demonstrate the effects of physical laws in the most simple but instructive manner; in this case, I shall not be using magic but shall rely on a physical effect, namely, inertia."

This preamble may persuade our audience to try for themselves, but all their attempts are likely to end in abysmal failure.

We, for our part, lift up the plate with the coins about a foot above the table, briskly drop it two-thirds of the way down and then quickly pull it toward us. The coins, deprived of their point of support by this maneuver, fall onto the table in a pile, retaining their original position.

Every one of our readers will be able to perform this experiment gracefully and with assurance after a few attempts, though, as we have said, some measure of agility is indispensable.

From Elbow to Hand

We take seven matching coins, which must not be too small, place them on top of one another, and bend our right forearm until it is horizontal, the hand held near our head. Now we transfer the pile of coins to our elbow and ask: "What is the quickest and simplest way of getting these coins into our right hand?"

While the spectators are still thinking about the problem, we swing our forearm suddenly forward until the whole arm is stretched out horizontally and – the seven coins are in our hand! The action has taken place at such lightning speed that the spectators still wonder how this feat was performed.

"It's simple – they traveled through the air," we explain. Clearly there could have been no sleight of hand, for if we had just produced a different set of coins, what had happened to the ones on our elbow? The audience would surely have heard them clatter onto the floor.

And, indeed, this trick, too, involves no magic, but is based once again on the application of very simple physical laws. Those who wish to master it had best begin by catching a single coin in the manner described. Through the sudden forward movement of the arm, the coin is flung forward in an arc; the hand, too, creates an arc, though a larger one and at greater speed. The two arcs intersect and if, after some practice, we manage to make the hand traverse its arc in the same time as the coin traverses its own, the coin will automatically end up in the hand, and all we have to do is to hold it.

The experiment may now be repeated with two coins on top of each other, and the reader will find that this is scarcely more difficult to perform than with a single one, since both coins receive the same impetus in the same direction and hence remain together. It is then only a matter of practice before the number of coins may be increased to seven or even more.

INERTIA AND THE CHECKERBOARD

INERTIA is reflected in the tendency of all bodies to preserve their state of rest or of uniform movement, in a straight line. Hence, if we wish a body to behave differently we must give it time to do so. That is why the railroad train driver does not start off at full steam and the car driver does not put a foot down hard on the gas pedal when starting off in the vehicle. Similarly, it is impossible to bring an automobile to a sudden stop, and only in the gravest emergency does the driver ram a foot on the brake as hard as possible. When we sit in a train, we can hear from the screech of the brakes that they are being applied long before the train enters the station. And when it finally does come to a

complete halt, it still does so too quickly for us to follow suit: the lower part of the body of anyone sitting in a seat is, it is true, so closely linked with the train that it comes to rest the moment the train stops, but the trunk does not yield immediately and swings forward, unless, by way of precaution, the passenger has taken a seat facing backward. And woe betide anyone who has risen from the seat before the train stops! For then, while your feet stand firm, the rest of your body sways forward and hence far too readily topples over.

On the checkerboard, too, this effect is easily demonstrated. Let us build a tower out of the counters, and knock one of the lower counters

out with a thin piece of wood, say a ruler or the lid of the checkers box, holding the wood in as near to a horizontal position as possible. The counter naturally moves because of the sudden impact. If the ruler is applied fairly slowly, the remaining counters have time to overcome their inertia and to follow the movement, with the result that the tower collapses. But if the counter is given a quick, hard, blow, it alone will be knocked out and the rest of the tower will remain in place.

And the Coin is Still There!

For another admirable demonstration of the effects of inertia, we need a piece of cardboard from which we very neatly cut out a square measuring some 5 × 5 inches, using a very sharp penknife and a ruler, and making sure that the square is as clean-cut as possible with no rough edges.

We now place the square on the tip of our left index finger, balance it, and place a fairly large coin on top. We should be able to tell when the center of the coin is right over the tip of our finger. Now we give the side of the cardboard nearest us a very sharp flick with the index or middle finger of the right hand. The cardboard will shoot into space while the coin, indifferent to the loss of its support, will come to rest on the tip of our finger. We can, of course, substitute other objects of the same sort of weight for the coin, or use other supports, or, indeed, carry out the experiment on a much larger scale.

And if you should fail at the first attempt, you may start with an easier variation in which the sudden flick-

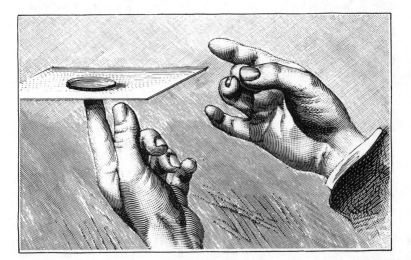

ing away of the cardboard can be dispensed with. Instead, you can snatch the cardboard away from the tip of the left index finger with a sudden jerk, having grasped it firmly between the thumb and index finger of the right hand; with this method, too, the coin will end up on the tip of the left index finger.

CUP-AND-BALL WITH DICE

OUR illustration shows how to prepare an experiment which is once again exceedingly simple to perform, but which is nevertheless likely to prove perplexing to those not in the know, for in this case, too, a trick is involved which will first have to be explained or fathomed out.

With our right hand we pick up a leather beaker, of the sort used in games of dice, together with two dice. The object of the game is to convey the two dice into the beaker one after the other. We must, therefore, toss the first die into the air and catch it in the beaker, which is easy enough to do with a little dexterity. It is a great deal more difficult, however, having tossed up the second die, held between two fingers, in the same way, to catch it again without making the first die jump out of the beaker; very rarely indeed will the beginner succeed in doing this.

To accomplish this feat, we do not toss the second die up, but allowing it to fall from our fingers, quickly move the beaker downward, whereupon both dice, deprived of their centers of gravity, adopt the initial velocity of their fall, that is, they move downward more slowly than our hand, the first die remaining within the confines of the beaker and the second responding readily to our wishes.

You see, dear reader, solutions are simple enough, provided we know how to apply physical forces so as to achieve the desired effects, and here, symbolically speaking, is the catch. It is best, therefore, if an experiment, like so many things in human life, is first tried out on a small scale, the better to sharpen our wits for what greater challenges we may have to meet.

EXPERIMENTS WITH CENTRIFUGAL EFFECTS

IF a body is swung in a circle, we refer to the force resulting from its attachment to the center of the circle as the centripetal force. Conversely, the body is affected by a force acting away from the center, which we call the centrifugal force. This force is the basis of the well-known experiment of tying a piece of string to a glass of water and swinging it around in a circle (see page 47). If the full glass is rotated with the correct initial velocity, not a drop of water will be spilled, because, since the centrifugal force is greater than the weight of the water, the water is forced to the bottom of the glass.

The spin drier is based on the same principle. It consists of a metal cylinder with a great many small

holes in its wall. If this container is filled with wet clothes and then rotated rapidly, the centrifugal effect will come into play: the wet articles of clothing are thrown against the perforated wall of the cylinder and the water forced out.

An attractive and surprising experiment which, moreover, does not require lengthy preparations, is shown in our illustration. Take a flat plate, place a napkin ring about half-an-inch high at its center, hold the plate firmly by the rim in the palms of both hands and toss it up, spinning it vigorously as you do so. It can easily be caught again even when, as in our illustration, the napkin ring ends up underneath, though still exactly in the center of the plate. If the axis of rotation is suddenly changed, the ring will be flung to one side. Should the reader have little faith in his own agility when performing this experiment, or should he be concerned about the safety of valuable crockery, he may use a cardboard disk instead, provided it is not too large; the experiment will be no less effective.

THE DEVIL ON TWO-STICKS

EVERY rotating body has a pivotal point or axis of rotation, that is, a point or rectilinear series of points that do not take part in the rotation. The axis of rotation may be held in position by a mechanical device as in the flywheel of a steam engine or it may be free as in a bicycle; indeed, it can simply be a mathematical concept without a physical presence, as in a rolling hoop or a spinning coin. All celestial bodies rotate about a free axis.

A well-known example of a free axis is provided by the spinning top in all its various forms. In it, as in all the last-named examples, inertia ensures that every particle rotating about the axis tends to cling to the plane of its orbit, and can only be forced out of it by an external force; this is known as the conservation of the plane of rotation. The top will not stand still as a result; on the contrary, it is always on the move; but the planes of rotation of all the particles are conserved, so that the axis of rotation, perpendicular to these particles, can only be displaced in a parallel direction. This is best shown by playing with the top in the air rather than on the table.

Such an aerial top is the so-called devil-on-two-sticks, or diabolo, a double tin cone used in conjunction with a string attached to two sticks. The devil is put on one of the sticks, the string is wound around the devil's waist, and the other stick is jerked away very quickly. As a result, the free axis will preserve its direction with a force that is the greater the faster the rotation and the larger the mass of the top. If the axis has been set at a slant from the start, the top will rise obliquely in a parabolic orbit and return obliquely in a continuation of that curve. The problem then is to catch it at its waist with the taut string in such a way that it rebounds into the air.

A Miniature Boomerang

In the hands of the Australian aborigine, the boomerang is an exceedingly dangerous weapon, one that he knows how to wield with deadly accuracy. If he is tracking game and wants to kill or stun it, he raises and aims the weapon, swings it back over his shoulder and, running hard, flings it at his quarry, which he rarely misses. The boomerang twists through the air like a screw, that is, rotates about its own axis, and deals its victim a mighty blow before falling to the ground. If it misses the target, it will move in a small upward arc and return to its starting point. A skilled hunter can send his missile in any direction he chooses, killing birds and small mammals at a distance of some two hundred paces – the cockatoo no less than the fleeing kangaroo in the bush. Against men, too, the boomerang can be a most perilous weapon, for we are unable to tell, as we see it approaching, which way it will turn and where it will strike.

In order to study this strange projectile in flight, let us construct a small cardboard model (see the illustration), taking care to make one side slightly heavier than the other. We now fit our small boomerang at an angle of 45° under the nail of our left index finger and flick it away with the thumb and index finger of the right hand. The projectile will rise and whirl through the air, finally coming to a stop but, instead of falling to the ground, it will then return to its starting point along the same flight path.

How can we explain these astonishing and unusual effects?

When we flick the boomerang, we impart a double movement to it, that is, a quick rotation and a general impetus; the rotation forces it to rise obliquely into the air and to conserve its plane of rotation until the impetus is exhausted. At this point it will still be turning but instead of rising higher, its weight will cause it to fall. Since, however, it also tends to conserve its plane of rotation, the resistance of the air will cause it to return, in a direction parallel to this plane, to the man who threw it in the first place.

A great deal of practice is required for success with a boomerang; first attempts are likely to end in failure.

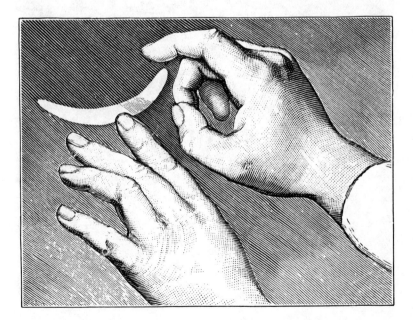

A Russian Soap–Bubble Carousel

For a change, let us now make a toy whose charm will gladden the eye. We take an eighteen-inch-long straw, straight and joint-free and bend it four times to obtain a rectangle measuring 6 × 2 inches.

Since the perimeter measures no more than $6 + 2 + 6 + 2 = 16$ inches, a remainder of two inches will protrude at one end – we make sure it is the thinner of the two. We now insert this remainder into the thicker

end, so that the rectangle is quite closed. With a second straw, we construct a rectangle of the same length (6 inches) but wider than the first by the thickness of two straws, and with a third, a rectangle just $1\frac{3}{4}$ inches in width. We now place the third rectangle inside the first, the second around the first, and arrange the three rectangles in such a way that they are placed at 60° from one another, thus constituting the diameters of a virtual hexagon. We have now made the wheel of our carousel. The frame, base and two struts are easily added. Wherever joints are needed we cut small slits or add a little glue. All that remains to be done is to bore through the vertical side sections of the frame with a thin, red-hot wire at intervals of six inches, and also through the centers of the longitudinal sides of our three rectangles; finally we push a wire shaped like a crank at one end

through the lateral sections of the frame and the rectangles of the "wheel." We now glue the three rectangles together, taking care to keep them at 60° from one another, and also glue them to the wire axle. We next cut out six small cardboard disks, which we attach loosely with pieces of thin wire to the six cross-sections of the wheel, drawing the ends of the pieces of wire through the centers of the disks. This done, we blow six soap bubbles about $1\frac{1}{2}$ inches in diameter and attach them to the disks, which we have previously moistened with soapy water.

If we turn the crank, we shall have before us the glorious spectacle of revolving soap bubbles, all shimmering with the iridescent colors of the rainbow, and shall be, moreover, in possession of a toy that affords us an excellent illustration of the effects of adhesion, interference and molecular force.

CHAIN INTO TOP

A FINE illustration of the effects of rotation is afforded by the experiment pictured on p. 46, for which we need a string some 16 to 20 inches in length, and a slender chain circlet measuring about ten to fourteen inches long.

We fasten the circlet to one end of the piece of string and hold the other end between thumbs and forefingers. If now we twist the string evenly, causing it to turn quickly (Fig. *A* of the illustration), the chain will gradually assume the shape of a horizontal circle (Fig. *B*), and as long as our efforts, and hence the

rotation, continue, it will maintain this circular shape.

An interesting variation is the following experiment: instead of the metal chain we attach a button to the string at a slight angle. If we again hold the string in the tips of our fingers and twist it, the button will rotate in its oblique position, but if we increase the force of the rotation the button will adopt a horizontal position. Simple though this effect may be, it is nevertheless put to practical use by engineers in the construction of various kinds of machinery.

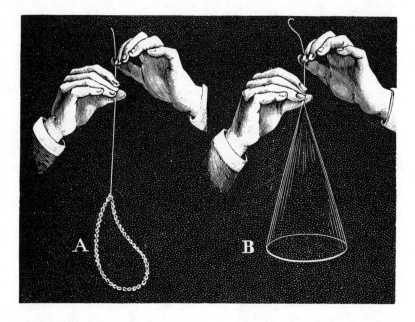

The Ascent of the Napkin Ring

Let us apply ourselves to centrifugal effects and rotation for the last time.

As a festive finale, so to speak, let us imagine a splendidly set table, as our illustration suggests, with an animated company, to whom we put the following question: "Do you believe that this napkin ring can rise into the air at my behest, and end up neatly around the neck of one of these bottles?"

While the assembled company are still puzzling over the answer, and we have surrendered to the pleasant state of luxuriating in our superior knowledge, the lady of the house may step up to the table with a smiling, "Why should the napkin ring not heed my behest as well?"

and with a quick rotation of her fore-finger stroke the inner wall of the ring, until the latter starts to rotate and rises up from the table. She then leads it up and above the neck of the bottle, stops moving her finger – and hey presto! the napkin ring drops down, no longer subjected to the impetus of the rotation, and lands around the neck of the bottle.

And we have humbly to confess that we have no monopoly when it comes to this particular Columbus's egg.

THE CIRCLING TUMBLER

It is in virtue of the centrifugal force which manifests itself with every rotation, that the rotating parts have a tendency to escape from the axis of rotation. The reader will know that Laplace and Kant based their views of the origin of the planetary systems on this very manifestation.

Where our planets now follow their paths, a glowing gas sphere of inconceivable size and temperature is said to have rotated at a moderate speed, the gaseous particles being kept in their places by a gravita-tional pull toward the center. Gradually, however, as this giant ball began to cool by radiation, the rotation quickened and the tendency of the gaseous particles to escape from the center increased, until finally they broke away from the gaseous sphere, and the first planet came into existence.

The following experiment allows us to follow this process on a small scale. Let us take a hoop of the kind children play with, place not too large an object, say a pebble or a

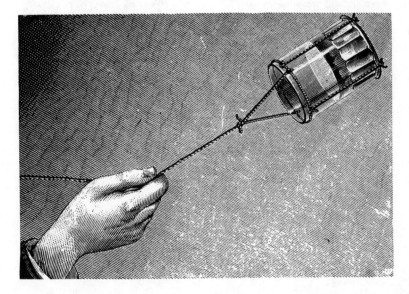

potato, on the inside edge at the bottom, and then rotate the hoop evenly with our arm outstretched. The pebble will follow the movement and remain in its original position. Even more surprising is the experiment illustrated above, and already mentioned in brief on page 45, in which the centrifugal force is made manifest by the fact that the water, in its attempt to escape from the center, is forced against the bottom of the tumbler.

For the performance of this experiment, the tumbler is stood on a cardboard disk and tied to it with string in such a way that it cannot fall off when the string is swung around a given central point. The circling motion which we impart to the filled tumbler with our hand or arm must, however, be executed with some skill and, most important, be at the proper speed from the start.

THE STRENGTH OF STRAW

By "strength," we mean the resistance a body puts up to the separation of its parts. This resistance will prevent tearing, breaking, crushing or twisting.

Our engineers and mechanics have precise knowledge of this phenomenon, so much so that they are able to tax no more than a fraction of the strength of the materials they use – from only a tenth to a quarter in the case of wood, and up to a third in the case of metals. Some of the most insignificant and delicate objects often have the greatest strength, for instance a cocoon or a cobweb. Thus a thread made by a spider, one square millimeter across, can take a weight of half a ton without breaking. Most people will also be surprised by the strength and carrying capacity of a straw.

To demonstrate this, let us ask for a carafe half-filled with water, bend a straw and put it in the carafe in the manner shown in our illustration. If we now offer to lift the carafe, which is fairly heavy, by the straw, we shall undoubtedly see a great many incredulous faces around us. But we can safely wager that, provided we do not tarry, the experiment will work every time. If the straw is left in the water too long, however, it will soften, and then we shall have to be prepared for broken glass.

Our audience will undoubtedly be astonished at the success of our experiment. Some of them will be struck by the bend in the straw, and may venture the opinion that the experiment is based on the lever principle. This is not the case, for the sole purpose of the bend is to ensure that the straw remains in the bottle: the experiment relies purely on the tensile strength of the straw.

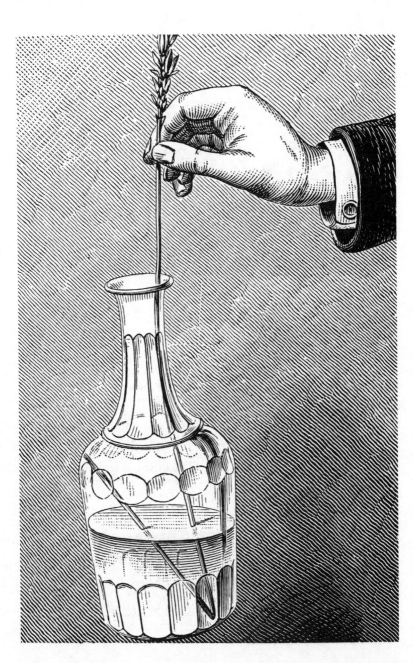

THE KNIFE AS NUTCRACKER

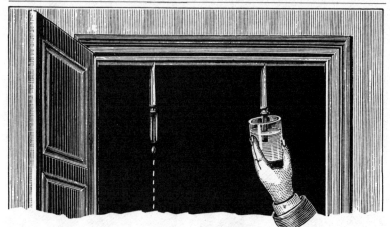

CAN an object, say a knife, dropped from the height of an ordinary door, attain sufficient momentum to crack a walnut?

People are sure to think that it cannot, and yet we can promise that it will work every time! Let us therefore, by way of a test, stick a sharp knife very gently, so that it just stays in place, into the molding at the top of a door frame – only where we are allowed to do so, needless to say – gauging the point at which the knife is likely to hit the floor as closely as possible, and placing the nut on that spot. Let us

then give the door frame a violent thump with our fist and the knife will come plummeting down – but it will usually hit the ground to one side of the right place. What must we do to make sure that it will hit the nut without fail? The answer is very simple: we dip the handle of the knife into a glass of water so that a single drop is suspended from it. The drop will not take long to fall to the floor, and we now place the nut where the drop landed. If we thump the door frame once again, the knife will fall down and unfailingly crack the nut.

A Blow with a Stick

Perhaps one or other of our readers has come across a trick that used to be performed at fairs – along with all the other mechanical diversions that served to amaze the public – and was performed as follows: the demonstrator would produce a thin wooden rod and invite two children, each holding an open razor, to take up positions facing each other at either end of the rod. Then a paper ring was hung over the blade of each razor and the ends of the rod placed in the two rings. The demonstrator now quickly gave the middle of the rod a heavy blow; the rod snapped, but the paper rings were neither torn nor cut by the razors.

The same experiment can also be performed in the manner shown in our illustration. Pins are stuck into the two ends of a stick, which we then rest on two wineglasses placed on two chairs. If we now give the middle of the stick a heavy blow, it will break in two but the glasses will

not suffer the slightest damage.

Occasional tricks of this fair-ground type, simple and showy though they may appear to be, often prove highly instructive. The explanation of the experiment just described is extremely simple. The descending cudgel hits the middle of the stick with great force and naturally the stick tends to move downward. This effect overcomes the resistance of the stick well before the disturbance is transmitted to its ends, and hence to the wineglasses. The result is spectacular: the center of the stick falls victim to the sudden influx of energy, while its ends are unaffected and remain in place.

THE TRANSMISSION OF AN IMPACT

WHAT do we understand by impact? It is the collision of at least two bodies, of which one (or more) must be in motion. Its effects are exceedingly brief but relatively strong. More precisely, an impact produces a compression and is propagated with a certain velocity. In order to demonstrate this second effect, that is, the transmission of the impact, there is a range of apparatus in most physics closets, including a row of ivory balls suspended at the same height, which is the most familiar of all. If we lift the ball at the end of the row and let it drop back onto its neighbor, that and all the adjoining balls will receive a blow but will not

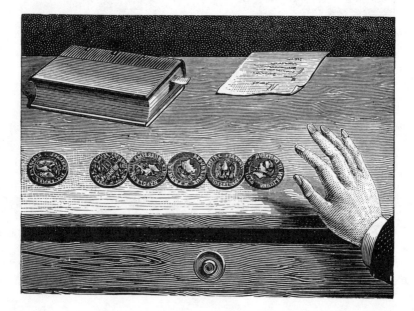

move from their places; they will nevertheless transmit the impetus toward the ball at the other end of the row, which will be repelled more or less strongly depending on the force of the impact. This apparatus can be reproduced with a little inventiveness and skill by the simplest of means. The quickest method is to reach into what we hope is an amply filled money box and to place a row of coins on the table, as shown in our picture. If we flick the first coin with the tip of our finger against the second, all the other coins will remain in place, except for the last, which will be repelled.

WATER INTO WINE

To change water into wine – which of us would not like to be able to do that on occasion! Our experiment calls for three wineglasses. Of these, two (*a* and *b*) must be as alike as possible; the third one (*c*) must be smaller and its rim, in particular, must have a smaller circumference. We place the first two glasses in the bottom of a bucket of water so that they are completely filled and no air bubbles adhere to them; to that end we moisten them inside and outside with wet fingers. Next we turn the glass *b* upside down in the bucket, and, placing it on top of glass *a*, remove the two glasses from the bucket and put them on a plate. After most of the water has run off from the outside of the glasses, we mop the rest off carefully with a soft cloth, so that the glasses are now quite dry on the outside but completely filled with water on the inside. We then place glass *c*, containing some red wine, on the upside-down foot of *b* and try to transfer the wine from *c* to *b* without touching either *a* or *b*.

The problem can be solved very simply: we place a wick (if need be a length of yarn will do) in glass *c* in

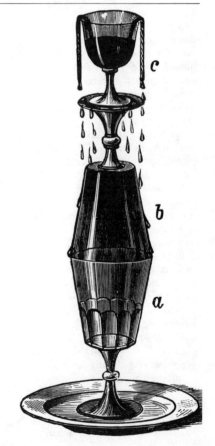

53

such a way that the middle of the wick lies at the bottom of the glass while its two ends hang down over the rim almost to the foot of *c*. We can now let nature take its course. Capillary action will cause the wine to rise into both ends of the wick and to drop down on glass *b*, the wick acting as a kind of syphon. Where *b* and *a* meet, the wine will, thanks to capillary action once again, seep into the water. Because of its less specific gravity, it will then rise in *b* and gradually displace the water, until it can be plainly seen to fill glass *b* entirely, and our task is completed.

WHICH IS HEAVIER?

OUR illustrations provide a graphic description of an attempt to determine the specific gravity of two liquids, namely water and wine. By specific gravity we refer to the relative density of a substance, or more precisely to the ratio of the mass of any volume of the substance to the mass of an equal volume of a standard substance. In the case of liquids and solids, the latter is usually water (at 39° F or 4°C), and with gases it is usually air (at standard atmospheric pressure, i.e. the pressure exerted by a column of mercury 760 mm high at a sea level temperature of 59°F or 15°C.)

But back to our experiment! We again take two identical wineglasses and fill one (1) with red wine and the other (2) with water. We cover the second glass with a sheet of writing paper and then place it upside down on the wine-filled glass (3). Now we pull the writing paper out very slightly (4), which requires great care lest we move the glasses; nor must the area of contact between the two liquids be too great. What happens next? Will the water and the wine blend with each other?

The reader will be inclined to answer in the affirmative. However, no such blending occurs; instead, the water, as the heavier of the two bodies, will

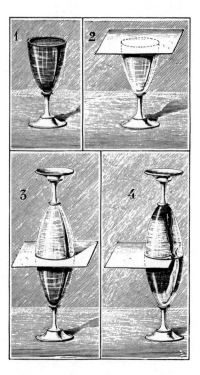

flow into the lower glass and gradually replace the wine, which will ascend, in the same proportion, into the upper glass. After a few minutes, the contents of the two vessels will have changed places: the glass originally filled with water will now contain the wine, and the glass originally filled with wine will contain the water.

THE JAPANESE PAPER FROG

THIS delightful object, made from folded green paper, comes from Japan, and although patience is required in its making, even the least nimble-fingered should have little difficulty in completing it.

First cut out a square piece of paper (Fig. 1), and fold it along the two diagonals *a*. Then turn it and fold it along the two lines *b*. When this has been done, it is a simple matter to make the shape shown on Fig. 2. Next fold the ends *b* and *a* as shown on Fig. 3. You will now have a series of eight small surfaces around the axis *oa*. Pick up the paper at *b* and fold it carefully so as to produce two new points, as shown on Fig. 4. Once you have repeated this operation with all eight surfaces of the folded paper, you will have made the shape shown on Fig. 5. Now you must fold every surface once again, turning the tip *s* toward the central axis (Fig. 6), and being particularly careful with the folds at the points. Fig 7 shows what has to be done to complete the frog. Taking the upper points *a*, make two creases to create the forelegs; the two lower tips *a* are used to form the rear legs. The left side of Fig. 7 shows the tips before they have been folded; the right side the completed legs.

A charming spectacle with which to delight your brothers, sisters and parents on a long winter evening!

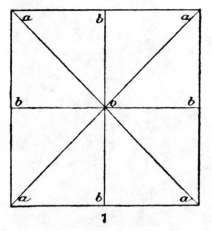

1

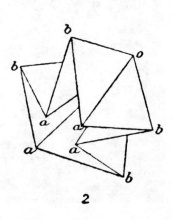

2

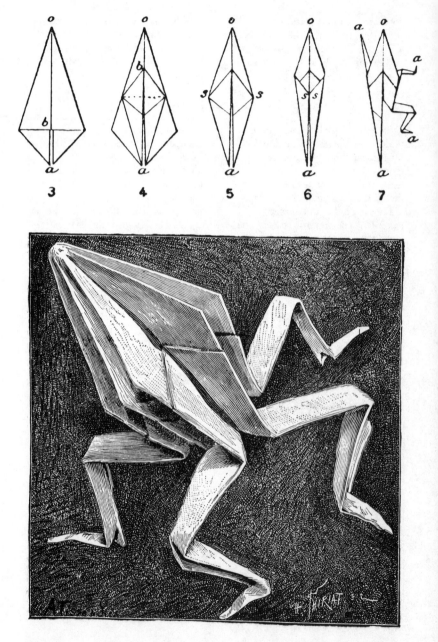

3   4   5   6   7

THE RUNAWAY COIN

ON suitable occasions, and with the help of a conical glass, such as a so-called champagne flute, with a relative small diameter at the lip, and two silver coins, say a dollar and a quarter, we can surprise our audience with a rather clever trick. We place the smaller coin at the bottom of the glass and the larger over it like a lid, as shown in our illustration. Now we undertake to remove the smaller coin from the glass without touching either the larger coin or the glass itself, boldly maintaining that it will obey our command. There will probably be some one amongst our friends who will, after some thought, volunteer to perform the experiment, and he may even hit upon the answer; he is much more likely, however, to give up after several vain attempts. Another may even go so far as to cast doubt on our magic powers, but we can soon put a stop to this impertinence, for all we need do to solve our task is to blow hard at the edge of the coin on top, which, as a result, will turn through a right angle. The downward-turning part of the coin exerts pressure on the air under the lower coin, which becomes so compressed both as a result and because of the blowing that the smaller coin is hurled in an arc out of the glass. Hardly has this happened than the larger coin reverts to its horizontal position and the "lid" is shut again.

THE LEAPING EGG

"WHAT will you give me," we were once asked by a friend, "if I remove a hard-boiled egg from this glass without touching it with my fingers or with any other object?" The friend, as our illustration shows, was sitting in front of a wineglass with the egg inside.

"You are well known as a conjurer," we replied with a smile. "But you can't possibly impress us with that sort of trick." "Why not?" "Because what you are going to do is so obvious. The trick is much too simple." "Oh, I see, you think that all I will do is pick up the glass by the stem and let the egg roll out. Nothing of the kind!" "Well then, let's see what you can do."

Our friend placed the egg inside the glass, making sure that about one third protruded from the top, puffed up his cheeks, began to blow into the glass and – hey presto! – with a graceful bound, the egg landed on the table.

"Well done, you have proved your point. We have to confess that we are utterly dumbfounded – the experiment itself is straightforward enough but the effect is quite astonishing."

"You'll be even more surprised," our friend declared with a smile, placing a second wineglass behind the first, "when I tell you that I can regulate the strength and direction of my breath in such a way that the egg will leap from the first glass into the second."

"Off you go, then."

He puffed up his cheeks again, began to blow and, with a daring leap, the egg duly jumped from the first glass into the second.

We were unsparing in our appreciation.

"Yes, it is a delightful experiment, isn't it?" said our friend. 'All it needs is a fair amount of practice."

A REMARKABLE TOP

CAN anyone make a top that can start to spin on its own? No? Then we must spring into the breach ourselves, for now that the question has been put, you, dear reader, will no doubt expect an answer.

All we need for our entertainment is a cork, a needle and a piece of paper cut into a square or rectangle. We place the cork on the table, stick the thick end of the needle into it, find the center of the piece of paper by drawing in the diagonals, and then balance the paper by putting the center we have thus determined over the point of the needle. First, however, we bend two opposite corners of the paper over, one downward and the other up.

Now we can start the experiment by putting our hand unobtrusively fairly close behind the top, as the illustration shows. It will not be long before it begins to move, and if we take our hand away it will come to a stop. Hence it only seemingly starts moving of its own accord. It is clear that the moving force emanates from our hand.

What is the explanation of this curious phenomenon? Is it magnetism? That answer is not totally implausible, but does not fit the case. What we have here is a very simple mechanical effect caused by the heating of the air. The fact that air can transmit its motion to other, solid bodies, we know from the sails of a windmill, from the sailboat, the air balloon and so on, but also from a great many toys, for instance, the spiraling top balanced on a pin over a warm stove. In our own experiment, the proximity of our hand, provided it is warm, produces a rising current of air which is caught beneath the downturned corner and starts rotating the paper, whose position on the point of the needle ensures that it is subject to the minimum of friction. The warmer the hand, the faster the rotation. Those, incidentally, who suffer from poor circulation cannot hope to perform this experiment easily, if at all.

ARTIFICIAL BIRDS

THE example of Icarus, who attached two wings to his shoulders with wax in order to fly toward the sun, found that they melted when he drew too near, and was thus badly let down by his invention, has been followed, despite his and numerous other failures, by many imitators.

Quite a stir was caused in 1786 by a young Frenchman whose flying machine was based on a similar principle. It consisted of a kind of yoke attached to his shoulders, with two movable and hinged shafts supporting the wings. The front wings were operated by hand, the rear wings by foot. Considerable success was claimed for the inventor – at the very least, that is, there were no reports that, like so many of his predecessors and successors, he had broken his neck.

A great deal of talk was also caused by a tailor from Ulm, who had to pay for his courageous flight with a cold bath in the Danube.

To those who are interested in aeronautics and mechanics we would like to commend the two little models of artificial birds depicted in our illustration. Their mechanism is simple enough, but extremely ingenious. The driving force of the beating wings is provided by a broad, tightly wound elastic band; the other details of the mechanism are clearly

shown in the illustration. At the top of the picture we can see a smaller artificial bird with lowered wings. The wingspan is 14 inches; the rubber band is 5 inches long and weighs about $\frac{1}{3}$ ounce; the total weight is about one ounce. The little machine becomes airborne just as soon as the elastic band is activated and released, and it can cover a distance of twenty yards and more.

The inventor also made a much larger mechanical bird, weighing no less than $1\frac{1}{2}$ pounds, which, with a head wind of twelve feet per second, covered a distance of just under 100 feet.

(The reader should bear in mind that these models, like all the games, tricks and experiments described in this book, were devised in the nineteenth century, that is, before the invention of the airplane!)

THE SPINNING COIN

CAN anyone start a coin spinning as fast as the wind about its own axis and then keep it going for some time?

"Nothing could be easier!" exclaims our neighbor at the table – with a superior expression. He produces a silver dollar from his pocket and stands it on edge, then, supporting the coin with the index finger of his left hand, he gives it a hard flick with the middle finger of his other hand. The dollar will spin round its axis so quickly that the eye cannot follow its motion, until gradually its speed decreases, and after a few wavering, circling motions it falls down flat on the table.

No one can accuse us of being bad sports, and we applaud him enthusiastically. But then we remark very meekly that we had been thinking of a quite different and slightly more complicated solution. Naturally, we shall be challenged to prove our greater brilliance, and as naturally we shall not wait to be asked twice. We ask our neighbor for the dollar he has just been using (but if its edge has worn smooth, we should find another coin with a well-milled edge), mark each end of one of its diameters carefully, place it on the table, lift it with two needles which we fit carefully to the exact points we have marked, bring the whole up to our mouth and blow at the top half of the coin. The pressure of the air will cause the coin to spin about its axis with great speed.

But how do we determine the diameter of the coin? To prepare for this we use a pair of compasses and a piece of paper to produce a circle that has approximately the same circumference as the coin, and draw a straight line through the center. If we now place the coin in the middle of this circle, it is a simple matter to mark the ends of the diameter on the edge of the coin.

The Floating Bread Pellet

Not so long ago, we had occasion to watch a well-known physicist enthrall a fairly large audience with a series of experiments, among which the following, simple though it is, attracted loud applause.

In this experiment, the end of a narrow metal tube is rendered airtight with whatever will do the job, and a hole measuring approximately $\frac{1}{16}$ inch in diameter is drilled about $\frac{1}{2}$ inch from the sealed end. If we now place a small pellet of cork or bread over the little hole, hold the tube up horizontally and blow continuously into the open end, the pellet will rise into the air from its position at rest and continue to float as long as the stream of air continues to flow. Should the reader try this experiment for himself he will find it highly effective. The floating pellet, incidentally, brings to mind fairground shooting galleries and those table tennis balls that are held three feet or more up in the air

by jets of water from a fountain, only to be made to drop down again by fluctuations in the force of the jets, thus offering the marksmen rather elusive targets while they dance up and down.

If there is no tube of metal close at hand, it should not be too difficult to substitute some other object, such as a small length of reed or bamboo; the effect will be no less spectacular.

THE FREELY SUSPENDED COIN

THERE are problems that can only be solved by those who have been let into the secret. One such is the problem of how to attach a small coin to a vertical wooden surface without resorting to any kind of adhesive or mechanical device. Those not in the know will try several

methods, unhappily meeting failure each time, while the expert will pick up the coin with the tips of his fingers, rub the coin energetically against the surface, then suddenly stop and push the coin hard against the wood – where it remains.

If we now ask our somewhat

perplexed spectators what physical effect may cause the coin to adhere to the surface, we shall probably be told that it is the force of adhesion. This answer, however, is far from correct, for though it seems that adhesion must be involved, it is not in fact. True, two smooth surfaces may adhere closely to each other after they have been pressed tightly together, and their adherence is the greater the larger the contact area. Thus well-polished silver and copper plates, placed on top of each other, will adhere so closely that they behave like a single solid. Similarly, we may find that our charcoal or pencil will stick to the paper on which we have drawn with them.

In our little scientific game, however, the imprint on the coin prevents such close contact. Instead, with the friction and the sudden pressure we have applied, heat almost totally displaces the air between its markings, with the result that the outside air now exerts enough pressure on the coin to make it stick to the wall. If this experiment should not succeed at the first attempt, it is advisable to change either to a coin that has a deeper impression, but that is also, perhaps, lighter, or to use a different surface, or perhaps to change both.

A New Version of the Magdeburg Hemispheres

Otto von Guericke's experiment with the famous Magdeburg hemispheres was intended to demonstrate the tremendous force of air pressure. To that end he used two sizeable hemispheres made of brass and copper, one of which was provided with a tube. Guericke fitted them together so that they formed a complete sphere and then evacuated the air inside with the help of the air pump he had invented. In the event, it took the combined power of thirty horses to tear the two hemispheres apart.

You will, of course, be familiar with this experiment, for you are bound to have studied it in your physics lessons at school; what you may not know, however, is that you can imitate this historically famous experiment with the help of two tumblers. Take two glasses of the same size, making sure that when placed rim to rim they fit together perfectly. Then stick a candle stump to the bottom of one of the glasses, place the glass on the table and ignite the bestower of light and warmth. You must now find a piece of strong paper, moisten it with water and place it on top of the glass. Then pick up the second glass and place it upside down on the paper, as the illustration shows. The adhesion of the two glasses, separated by the paper, must be perfect if the experiment is to succeed. The candle will soon go out, the air in the lower glass having become considerably rarefied. If you now pick up the top glass, the lower glass will remain attached to it and be lifted up as well. The pressure of the atmosphere outside holds the two tumblers firmly together, just as it did with the two Magdeburg hemispheres in the classical experiment.

There will be times, admittedly, that the paper between the two glasses will burst, but even then the experiment will succeed, provided all the other arrangements have been made as we have described them here.

BLOWING A CANDLE OUT WITH A BOTTLE

THE heading above appears to promise a very curious experiment. To perform it, we use our right hand to pick up an ordinary wine bottle by its base. We close the opening, which is just under an inch across, with the ball of our left hand, but in such a way that we can produce a small aperture by lifting our hand very slightly. This aperture we now

cover with our mouth, and then we blow very hard into the bottle for some three to four seconds, quickly closing the aperture again with the ball of our hand. As a result, the air in the bottle becomes compressed. All that needs to be done now is to turn the bottle upside down, as shown in the illustration below, and bring it up to within about an inch of the flame of a burning candle, which will be blown out by the fairly strong gust of air released as soon as we once again move our hand very slightly away from the neck of the bottle.

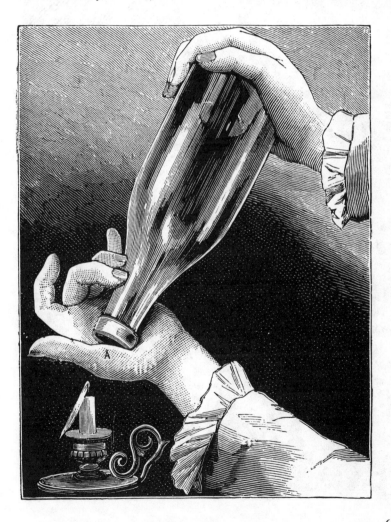

A PASSENGER-FERRYING SOAP BUBBLE

EACH of our readers has no doubt blown soap bubbles, and is aware that they float upward because they are filled with breath from our lungs which, because of its higher temperature is lighter than the air outside. If the right equipment is to hand, it is possible, of course, to fill the soap bubbles with hydrogen, in which case they need to have a diameter of only 2 inches to rise quickly to the ceiling. Experiments like these led us to the idea of making the largest possible soap bubble and attaching the figure of a tiny balloonist to it. This is not something we achieved without difficulty, but anyone can do it if the experiment is performed carefully. To that end, you should cut a manikin one inch tall from a piece of tissue paper, tie it to one end of a slender length of thread, to the other end of which you must attach a $\frac{1}{4}$-inch disk

of fine, white writing paper. The thread is tied to the paper manikin and the small paper disk by simple knots. The disk can be made to stick to the soap bubble, as shown in our illustration on p. 70, by simple but careful contact. If you now slightly vibrate the soap bubble pipe, the bubble will break off and, like a balloon with its balloonist, rise gaily into the air (see p. 71), causing great mirth to any spectators.

FORCING AN EGG INTO A BOTTLE

IT may easily be demonstrated, even without an air pump, that dense air exerts very much greater pressure than rarefied air. To prove this, we ask the lady of the house for a hard-boiled egg, and with her permission fetch an ordinary carafe from the kitchen or the sideboard. We carefully peel the first, and then make a paper spill. This latter we set on fire, and throw it burning into the bottle. What happens next? The air in the bottle is expanded by the heat and partly escapes, so that within a fairly short space of time only a small amount of rarefied and heated air is left inside the bottle. We now quickly place the peeled egg, which we have kept close at hand, on the neck of the bottle, like a cork, but do not push it down. What ensues? The air in the bottle, since the spill will have finished burning fairly quickly and will no longer be giving off heat, will gradually cool and at the same time become denser. As a result the air pressure in the bottle will be greatly decreased and give rise to a most peculiar phenomenon, and an ultimate catastrophe. For the egg, yielding to atmospheric pressure, will gradually travel down the neck of the bottle, hugging it tightly and elongating like a snake, sliding down ever more quickly until finally it ends up with a thud at the bottom of the bottle.

Although we did not attempt it ourselves, this experiment can doubtless be performed with other objects, say a soft rubber ball, a suitably sized cooked potato, large plums or similar fruit; enterprising readers are encouraged to try for themselves.

AN INEXPENSIVE TROMBONE

IT is well known that bird song can be imitated most effectively by rubbing a glass tube along its length with a piece of soft cork. Exactly the same effect can also be attained if the cork is rubbed along an ordinary glass bottle. The resulting sounds will vary with the speed of the action. If, instead, we take a glass tube a foot long with a diameter of just under an inch, and expand one end slightly over a spirit flame to make a mouthpiece, we can turn it into a musical instrument whose tone is reminiscent of that of a trombone. To do this we roll a piece of cardboard around the tube and glue its edges together so that the length of the tube is doubled. This double tube is long enough to make a good trombone with a low pitch. By moving the cardboard tube up the

glass tube, we decrease the length of the vibrating column of air, and gradually raise the pitch as we proceed.

This type of instrument resembles – we make bold to say – the antique sackbut. It is not at all difficult to play: the layman will be able to divert himself with it and, if he perseveres, become an accomplished artiste.

We know a young man, whose portrait the reader will find on the facing page, who plays this primitive instrument with such skill that people shake their heads in wonder at the purity and perfection of the notes he produces.

The Bottle-Harmonica at Concert Pitch

To create the bottle-harmonica (or musical bottles) shown in the illustration we must take two ordinary chairs and rest a wooden pole – or better still a bamboo stick – on the seats and a second pole on the backs. We then hang any number of wine bottles from the poles by tying lengths of fine twine around their necks and attaching them to the poles at equal intervals. The empty bottles may now be tuned by pouring water into them. We will not specify the precise amounts of water needed for any bottle, though it would not be too difficult to do so, preferring to leave this task to the musical talents of the performers themselves;

here, too, an ounce of practice is worth more than a pound of theory!

The instrument is played with long-handled wooden hammers, similar in appearance to croquet mallets and, because of its ample size, it can be used by two or even three people at once. As for the best technique, we again reserve our judgment. Our own experience limits us to the observation that well-trained musicians may safely be allowed to give a concert in an enclosed space, but that if they want to practice they had best take to the open air, if possible in a remote part of the garden, lest they strain delicate nerves beyond endurance.

THE MELODIOUS DULCIMER

To conclude our gallery of improvised instruments, we should like to show you, dear reader, how to construct a dulcimer, an instrument on which it is possible to play delightful and charming pieces of music and which, at the same time, helps us to study the laws governing the vibration of strings. The multi-plicity of these laws offers much to interest us, since strings can be actuated in conjunction with a sounding board or by plucking, striking, pulling with a plectrum or by blowing. In all these cases, it is never one simple vibration that is produced but, depending on the manner and place of the actuation, a

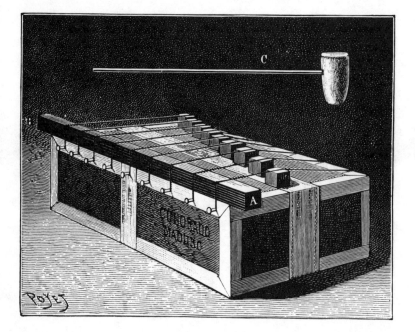

whole series of harmonic overtones in addition to the fundamental. To enter into the subject in any further detail would take us too far afield.

To construct our musical instrument we first need an ordinary, clean and undamaged cigar box and, as the illustration on the facing page shows, two rows of small nails which are knocked in at regular intervals close to the top edges of the box. We now stretch fine brass wires between each pair of nails, and insert a carefully planed square measuring rod (*AB*) under them on the left-hand side and then, at a slant, a line of small wooden dice, which we push to and fro until the strings sound the notes of the scale. If we wish, the outside of the little box can now be prettily painted, or, according to taste, given a light or dark stain, **and** our dulcimer is finished. All we need now is to add two small strips of whalebone; one end of each is stuck into a cork that has been carefully cut and filed down until it is shaped as shown at *C*. With these two little hammers reader can now try their skill and – let us hope! – after some practice produce truly delightful results.

WELL GRUNTED!

IF we are amongst close friends or within our family circle, and if there is a volunteer present who will not take a somewhat crude joke amiss, we can perform the following "experiment." The demonstrator puts a wineglass in our plucky volunteer's mouth, as the first illustration shows, and wraps his head in a napkin, saying:

"I shall be asking our esteemed friend to be kind enough to make a series of loud sounds, while I, for my part, will be holding the napkin tightly at the back of his head so that all the sound waves can impinge upon it. At the same time I would ask the audience to kindly observe the most scrupulous silence. The napkin is now around your head, so you may start producing the sound waves. Make prolonged, and deep sounds, for instance an *o* or a *u*, as well as a shrill *e*, and try to achieve a really good effect."

While the volunteer is grunting and squeaking away, the demonstrator ties a knot in the napkin at the back of the volunteer's head, so that the corners hang down on either side like ears, and then gently touches the white cloth at two spots in the region of the eyes with a cork blackened over a candle flame.

"Please continue, if you will, with your sound waves. Now, ladies and gentlemen, can you all make out the sound wave picture? Does it not bear the most striking affinity to the sounds you are hearing? We could call it the grunting picture, don't you think? But what are you doing, ladies and gentlemen, you must not laugh, or the whole effect will be lost. I really must ask you. . . ."

But to no avail, laughter resounds from all sides, drowning out the gruntings of the surprised volunteer, who soon tears the napkin from his face to discover to his indignation that he alone does not know the reason for all the mirth.

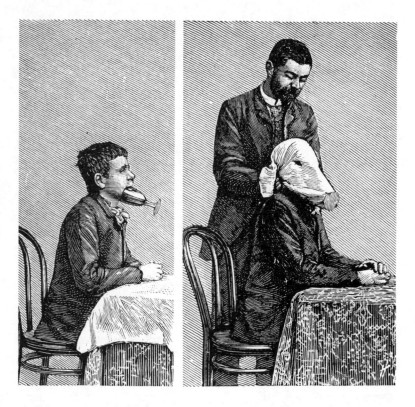

IMITATION THUNDER

THE sound of claps and prolonged
rolls of thunder can be produced
in an easy but startling way with a
piece of string. We ask someone –
who does not have to be a child, as
illustrated on page 79, for the
experiment is unlikely to fail to
impress even adults – to put his
hands over his ears, then we pick up
the string (which had best be fairly
thin and have tightly twisted strands)
and wind it around the head of the
volunteer so that it runs across the
middle of his hands and is tautly
stretched. Holding the loose ends
of the string in one hand, we now
pluck it like a violin. If we use our
fingernail, thunderclaps or loud
rumblings will be heard. We can cut
the sound short at will by squeezing
the string between thumb and fore-
finger. Cannon fire, too, can be
imitated in this way. The longer the
string the louder will be the effect.
  Most readers will know how to
imitate the sound of bells, but for

those who are not privy to the secret, we shall briefly describe this no less surprising experiment. We suspend an iron rod, some three feet long, say a poker, from a doubled piece of string, and wind the two ends of the string around each of our fore-fingers, which we then use to stop up our ears. If the poker is now set swinging and allowed to strike some object, we shall hear a sound reminiscent of, but much more beautiful than, that of a great bell.

VISIBLE VIBRATIONS

A FUNDAMENTAL law of the science of acoustics goes as follows: every body producing a sound is in a state of vibration; by the term vibration we understand the regularly repeated to and fro movements of the source of the sound. We distinguish between noises and sounds. The former make an irregular, varying impression, as, for instance, the roaring of a waterfall, the babbling of a brook, rattlings, rumblings and so on. In the case of sounds, by contrast, the listener has a very definite, continuous sensation that can be described as periodic, that is, repeated at regular intervals. The underlying mechanical process can be easily exemplified with the help of a very simple experiment.

A wineglass is turned upside down and a small pendulum, made

of a piece of string and a shoe button begged from a shoemaker or taken from a discarded boot, is fastened to the stem of the glass. The button must hang as far down the side of the glass as the illustration on page 80 shows. Once these preparations have been completed, we hold the glass up by the base between the tips of the fingers of one hand, and tap the side of the glass with a pencil lightly held in the other hand. The glass will give off a sound, and while this sound continues the button will perform hops along the glass wall, thus illustrating the vibrations.

THE WATER CANDLESTICK

WHEN we returned home after dark one night, and struck a match in the bedroom to light a candle, we discovered to our dismay that the candlestick was missing. The night before, the candle had burned down and the candlestick had been removed for a new candle to be fitted. Would we have to retire to bed in the dark, possibly breaking some precious object while groping about in the room? But wait a minute, the candle stub was still lying on the bedside table. Quickly, before the match burned out, we lit its wick.

But we were now in the rather unenviable position of holding the burning stub between our fingers, while we looked around for somewhere safe to put it. Then we were

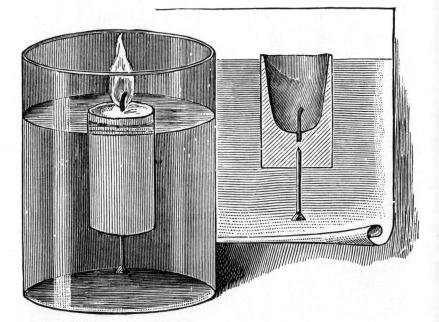

suddenly put in mind of the night light that used to burn in our nursery, and we quickly set to work. We picked up a nail, heated its tip in the candle flame and pushed it into the bottom of the candle stub. Next we poured water into a glass and introduced the weighted stub into the water. We had chosen a nail of exactly the right weight: the candle stub sank so that its top was just level with the water surface. As the candle kept burning, the stub grew shorter, but it also grew lighter and rose, so that we did not have to fear its premature extinction.

And so everything went according to plan. However, another phenomenon greatly surprised us: the candle burned hollow, that is, its edge stayed up like a rim, and it soon looked as if the flame was screened by a ground glass cylinder. After some time, we hit upon the explanation that at first seemed to elude us: the heating effect of the flame was opposed by the cooling effect of the water, and at the edge of the candle the latter was the greater of the two. The strange manner in which the combustible material was being consumed could be explained in no other way.

We stayed up watching the water candlestick for some time and, because of this quite ordinary but nevertheless fascinating discovery, we nearly forgot to retire to bed.

MUSLIN AND A GLOWING COAL

THOSE bodies which very rapidly absorb the heat of others and distribute it quickly throughout their mass are called good conductors of heat; by contrast, those bodies in which this process takes place very gradually are known as bad conductors. Thus, if we heat one end of a length of wire over a flame the other end will very quickly become so hot that anyone foolhardy enough to pick it up will drop it almost immediately. This does not, of course, happen with a piece of wood, which we can hold in our hand until the flame almost reaches our fingers.

Few people will believe, however, that it is possible to touch a piece of muslin with a glowing coal and leave the cloth intact. And yet it can be done, as the following experiment will demonstrate.

We take a copper sphere some 4–5 inches in diameter and cover it with a piece of muslin or fine linen, pulling the cloth tightly over the sphere and holding it at the bottom so that the sphere is supported by it. Then we pick up a piece of coal from the fire with a pair of tongs, place it on the upper pole of the sphere, and blow on the coal until it glows. Although the coal is now red hot the muslin does not burn, the reason being that the copper, which of all the metals is the best conductor of heat, absorbs all the heat so quickly that none can go into the cloth. An experiment that surprises quite as much as it adduces a convincing demonstration of the conduction of heat!

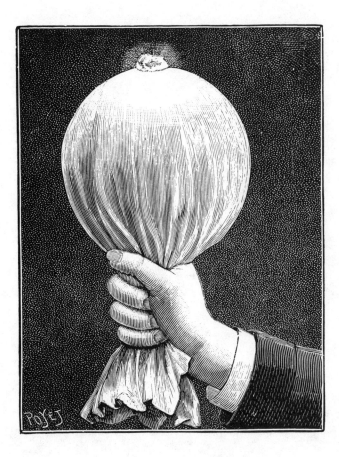

THE ELECTRIC WALNUT

How often has not some trick, some magical feat, been explained as a result of the powers of electricity, although this particular natural force has no connection whatever with it? For experience teaches us that the applause is the greater the more improbable the whole thing sounds.

An electric walnut is something altogether novel, and however unlikely it may sound it is produced easily enough. Simply rub the blunt end of a walnut – which is being held, pointed end upward, between thumb and middle finger, with the index finger lying on top – on your coatsleeve or some woolen cloth.

After it has been rubbed for a time, the audience will be astonished to see that it remains hanging from the index finger, and can only be freed with a slight tug.

If you now ask someone else to repeat the experiment, he is sure to fail. Why? Because he does not know the trick that is needed to succeed with this little bit of hocus-pocus!

The rubbing of the nut serves no purpose whatsoever, and merely helps to confuse those not in the know. During the rubbing, however, the middle finger and the thumb must exert strong pressure on the seam of the nut, so that it opens up at the top and the skin of the inddx finger, which is being pushed down-ward, is pinched into the resulting slit; the nut will then cling to the finger as soon as the pressure on its walls ceases. Now we can hold it up triumphantly and with due solemnity before the gaze of the astonished audience, who needless to say, will be full of admiration for the illusionist.

This trick ought not to be repeated too often, if the secret – elasticity, instead of electricity – is not to be discovered. The rubbing on the cloth, of course, must be carried out with an air of great importance. The opening of the nut may be assisted beforehand by a preliminary splitting of the seam with a thin blade.

THE ELECTRIC BROOMSTICK

IF we suspend a pith ball from an insulating silk thread (best of all is a thread taken from a white silk ribbon) and bring it close to a glass rod that has been rubbed with silk, it will be attracted, only to be immediately repelled. If we then touch the ball with our hand, it will be attracted all over again, only to be repelled once more. Clearly the pith ball has received an electric charge from the glass rod, and retains that charge because it is suspended from a silk thread. When it is touched with the hand, it surrenders its charge, returns to its original state, is attracted all over again – that is, recharged – and then repelled. Such an electrified ball is attracted by everybody not carrying an electric charge, for instance by a nearby metal object – the clearest sign that it does indeed carry a charge. If we now suspend two such pith balls and electrify one with a glass rod rubbed with silk, and the other with a rod of ebonite rubbed with fur, then the ball repelled by the glass will be attracted by the ebonite and the one repelled by the ebonite will be attracted by the glass.

A similar experiment – do not be alarmed – can be performed with a broomstick. Balance the latter, as the illustration shows, on a table, so that half of it protrudes over the edge and can be set swinging by anyone blowing at it. Then take a thick stick of sealing wax, rub it fairly hard against anything woolen, for instance a coatsleeve, hold it close to the upper or lower end of the broomstick, and the latter, without having been actually touched by the sealing wax, will begin to swing noticeably. This surprising effect is based on the electric charges present and can, in this case, serve to demonstrate that electricity is not confined to certain bodies, but that all bodies have what is known as latent electricity.

ANOTHER WAY OF FISHING

THE toy is depicted in our illustration will give the little ones a great deal of pleasure. First we must assemble the fishing tackle: the rod is a stick 12 to 18 inches long, the line is borrowed from the sewing basket, the hook is fashioned from a pin (see insert A in the illustration) and a small ball of sealing wax. If we can shape the latter to look like a worm serving as bait, the appeal of the game is sure to increase. Next we cut small fish shapes out of thin paper and use crayons to color in the eyes, scales and fins. If we are skillful enough

we can add a square cardboard box without its base or lid – that is, open at the top and at the bottom – to serve as an aquarium, coloring the sides suitably.

For the game itself, we lay the fish down on the table or in the aquarium and define the rules of play. The winner is the one who catches the largest number of fish first, and he is awarded some sort of prize. It goes without saying that each contestant must be provided with a fishing rod. We need not go into detail as to the way the fish are caught, for the reader is well aware that the sealing

wax can be electrified by friction and will therefore be capable of attracting various objects. The anglers can, however, also make use of magnetic effects, in which case they must attach small horseshoe magnets to the ends of the fishing lines; the fish must be cut out of heavier paper and small pieces of

iron inserted in their mouths. The equipment gains in durability in this way and may be brought out for use during idle hours time and again.

There is no need to say, "Enjoy yourselves," with this game, for experience has taught us that no one can fail to derive great amusement from it.

THE ELECTRIC DANCERS

ELECTRICITY, which plays so important a role in industry and the modern home, can, in its simplest manifestations, serve for many diverting experiments and games. To prove this claim, we use an ordinary sheet of glass measuring some 15 × 18 inches and place it between two fairly thick volumes, as our illustration depicts; the glass

should be about 2 inches above the table top. Next we cut out a number of small figures with scissors: ladies, gentlemen, harlequins, frogs, imps and so on, which should not be taller than about $\frac{3}{4}$ inch, a little smaller than the figures shown at the top of our illustration. They are best cut from sheets of paper of different color, which will give them

a very pleasing appearance.

We now place the little figures between the sheet of glass and the table, where they can be laid down flat next to each other in any order we like. Then we take a woolen or, better still, a silk cloth, crumple it into the shape of a ball, warm it slightly and then rub the top of the glass very hard with it. We shall be able to observe immediately how the electricity produced in this way attracts the paper figures, for they will suddenly stand upright and leap up to the glass ceiling of their little ballroom, only to be repelled and to fall back down onto the table, there to begin their lively and comical dance all over again. If we stop rubbing the glass, the merry proceedings will continue for a little while, and once the dance has finally come to an end, a slight touch of the hand on the glass is sufficient to revive the little figures once more.

THE INDUCTION TOP

WHENEVER metal is moved in the vicinity of a magnet – inside a magnetic field, as it is usually called – the magnetic field will tend to arrest the movement. This effect has been applied most ingeniously to the improvement of various scientific measuring instruments, but here we shall merely describe a very simple experiment, intended to persuade the reader of the existence of the peculiar effect known as induction.

To that end we spin a top, consisting of a disk of soft iron, in the usual manner with a piece of string. When the top is at rest, it is attracted by a horseshoe magnet placed in its vicinity; however, if the top is spinning and the magnet is brought close to it in such a way that the shanks of the magnet are vertical to the plane of the disk, the top will

begin to dip away, and its inclination will be the greater the stronger and closer the magnet. Once the speed of the rotation has dropped below a certain threashhold, however, the attractive force gradually predominates, so much so that the top finally attaches itself to the magnet.

The explanation of this curious phenomenon is simple enough. When the disk rotates with great speed in the vicinity of the magnet, it becomes the seat of induction currents. These currents and the magnet repel each other, and the strength of the repulsion decreases with the decreasing speed of the rotation. When the speed dwindles to below a certain point, repulsion makes way for attraction. That induction currents are indeed responsible for this phenomenon is shown by the fact that the rotating top will be attracted by the magnet when the shanks of the latter are parallel and close to the edge of the former. In that case, for reasons whose elucidation would take us too far afield, no induction currents are set up in the disk.

MAGNETIZING A KNIFE

A STEEL rod may be magnetized merely by being stroked with a powerful magnet. That half of the rod which is destined to become the north pole is stroked some ten times on either side with the south pole of the magnet, the motion being started firmly at the center of the rod and continued with moderate speed beyond the end before the magnet is returned in an arc to the center. Similarly, the half intended as the south pole is stroked with the north pole of the magnet. It is helpful to attach the rod to the table by a small iron band fitted over its exact center. More powerful magnets can be made if the opposite poles of two magnets of equal or near equal strength are placed over the center of the rod in such a manner that each forms an angle of 30° with its half of the rod, both magnets then being guided at the same speed along the rod and beyond. In this case, too, it is best to return them to the center of the rod by making an arc through the air, care being taken that the two magnets never touch.

In what follows we shall describe how a table knife can be rendered magnetic, proceeding as shown in the illustration on page 91. The blade of the knife is placed on the underside of a coal shovel, and is then stroked with the top of a closed pair of tongs, always in the same direction, from the handle to the tip of the blade, the knife being turned over from time to time so that either side is thoroughly stroked. If this operation is continued for some forty to fifty seconds, the blade will be magnetized, and will now lift a needle or a steel nib with great ease.

The magnetism produced in this way persists for a long time. This simple process is not usually described in textbooks of physics, and yet is interesting enough to invite study. Closer investigation will show that the tip of the blade has become the north pole.

A Leyden Jar and an Electrical Tea Trolley

Making a Leyden jar is not at all difficult. We procure a cylindrical glass, as shown in the illustration, warm it and line it, inside and out, base and sides, with aluminum foil. If there are bumps or bubbles in the foil that prove impossible to remove by rubbing, we must make a clean cut with a sharp knife and rub again until the foil is quite smooth. The work is started on the inside of the glass, the base being lined first and care being taken that the foil goes right up to the inside wall of the glass (making several cuts, if necessary!). The sides are then dealt with in the same way, as is the outside of the glass, an uncovered edge being left around the top 2–4 inches deep, depending on the size of the glass. Then we lay a cardboard disk, of exactly the same size as the base, over the bottom of the glass, and finally provide a wooden lid for the jar. We drill a hole through the center of both the lid and the disk, and thread a metal wire, rod, or tube, which must be some 3 inches taller than the glass container, through it. We now add a metal sphere to the upper end of the rod, and a bunch of aluminum foil to the lower end, thus bringing the rod into contact with the inner lining of the glass – and our Leyden jar is ready.

The reader might well think that nothing could be simpler, and yet we have devised a formula that achieves the same end in a matter of seconds. We simply take a tumbler, fill it half full with lead shot, stick a silver spoon in it, and – we have our Leyden jar.

We can also make a very simple generator for charging our Leyden jar. Taking a lacquered tinplate tray, some 12–15 inches long, we place it on top of two wineglasses. Then we cut a rectangle out of cardboard to which we glue two paper handles, and if we now place our construction on a warm cooker to dry out, then lay it on the table and brush it hard with a stiff clothes brush, the paper will soon be strongly electrified. Lifting it off the table, we place it on the tray and touch its edge; we can now draw a spark half-an-inch long from the tray with one of our knuckles. If we use a coffeespoon to draw the charge instead, transferring the latter to our Leyden jar and repeating the process several times, the charge in the jar can become so great that when we discharge the jar the sensation may be most unpleasant.

THERMOELECTRIC CURRENTS

A LOVELY experiment! It is true that one has to go to the trouble of making the necessary apparatus, but this should give our readers few headaches.

To begin, then. The apparatus consists chiefly of a many-spoked wheel, made of various materials: the rim of the wheel is an alloy of nickel and copper; the spokes are copper wire; the hub is a small plate of sheet copper; in addition we need some soldering solution and tin. We use a cylindrical object to shape the rim of the wheel before soldering it together; once this is done we attach

the spokes by bending their ends around the rim and then soldering each in turn. The copper plate serving as the hub is soldered to a single spoke; the other spokes merely serve as supports. The hub is shaped like a dish, with the concave side turned downward. We make a small indentation in its center to receive the needle that will carry the whole apparatus. The needle is pushed into a cork which, in turn, is inserted into a candlestick. To make sure that the apparatus will stay in a horizontal position, we hang small cardboard pennants (four will do) from the spokes. If we now heat any one of the soldered spots on the rim of the wheel with a candle flame and hold a horseshoe magnet on the opposite side so that it lies over and under the plane of the wheel (see the illustration), the wheel will be set in motion. The finer the copper wire, the faster will be the motion. (For a wheel with a diameter of 4 inches it is best to use 0.05- to 0.075-inch wire.) The experiment represents the transformation of thermal into electrical and ultimately into mechanical energy. Through the heating of the soldered rods, an electric current is set up which is communicated toward the opposite soldering point and the rim of the wheel. Once there it forks back and returns to the first soldering point. Without the help of the magnet nothing would happen, because the two effects would cancel out; with the magnet, however, rotation will always take place.

DRAWING WITH FIRE

APPARENTLY blank sheets of paper can provide amazing entertainment by revealing singed figures, such as that shown in the illustration. The preparations are so simple that a brief description should be enough to allow anyone to produce a "fire picture" for himself – provided he can draw a little.

We dissolve saltpeter in cold water until the solution is saturated, which we can recognize by the fact that, despite vigorous stirring, some salt particles remain at the bottom of the glass. Using this solution and a pointed paintbrush, we draw the outline of an animal or some object on blank sheets of thin newspaper, and allow the paper to dry thoroughly. When dry, the drawing is completely invisible, enabling us, if we should so wish and if we are in the right company, to give the experiment the semblance of a little bit of magic. First we light a match, blow out the flame and with the still glowing tip touch a carefully chosen spot on the drawing – or rather, on the apparently blank paper. The saltpeter will catch fire immediately, and the flame will faithfully follow the lines that have been drawn with the solution, until finally the figure is burned out of the paper.

This experiment is not entirely unknown, but it is bound to afford great amusement, especially if the drawing is of a jocular nature.

CRYSTALS ON A THREAD

A MOST remarkable crystallization experiment may be performed by dissolving as much soda in water as the latter can absorb. When the crystals no longer vanish, despite repeated stirring, we can be certain that the solution is saturated. We then decant the clear liquid carefully into a second vessel, take a thread and tie one end to a red chili bean and the other to a supporting wire, glass rod or match. In the same way, we attach a nonporous body, for instance a pebble or a piece of glass, to our crossbar, which we then place across the top of the glass vessel containing the liquid, so that both

bodies are fully immersed, as our illustration shows. If we now leave the whole arrangement to its own devices for a while, an amazing crystal structure will begin to grow on the bean: spiky needles of soda attach themselves to it and soon cover it completely so that it looks like a small hedgehog, the bean itself being completely hidden from sight. By contrast, the body attached to the left side of the carrier (in our illustration, a small, round stone of the kind children use in their games) does not alter its appearance in any way.

The cause of the crystal formation

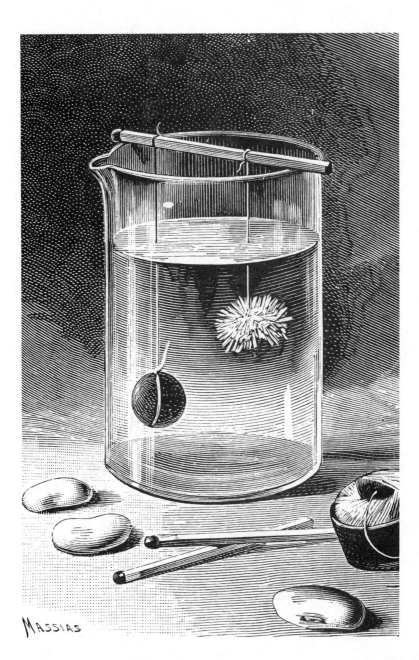

around the bean is not difficult to find. Unlike the round stone, the bean is highly porous and hence hygroscopic, that is, it absorbs water and swells. But only pure water is absorbed; the dissolved soda is rejected and attaches itself to the outside of the bean in the form of crystal needles; the greater the power the body has to absorb water the more pronounced the crystal formation. The stone, on the other hand, is not porous, and hence neither absorbs water nor rejects the soda. This experiment, easily performed, will gladden everyone's eye.

SWIM, LITTLE FISH!

WE cut a little fish, about 2 inches long, from a sheet of ordinary writing paper, not forgetting the caudal, dorsal and ventral fins (it is advisable to sketch the outlines in beforehand). Not content with this, we seize the opportunity to show off our artistic talent, too, reaching for our crayons and giving the little fish a scaly coat as true to nature as

we can make it, keeping the back a dark grayish brown and the abdomen slightly lighter: yellowish brown shading off into violet and then white. If our fish is to be a trout, we must not forget to scatter black spots all over its body. When we are satisfied with our handiwork, we use a pair of scissors to cut a narrow strip from the tail to the middle (see the illustration) where we cut out a circle. Now we fill a shallow dish with water, and place the fish carefully on the surface of the water, so that the upper side of the paper remains absolutely dry. Our next objective is to persuade our little fish to swim across the dish. How do we do that? By blowing, for instance? No, we have something

much better, much more interesting, up our sleeve. We take a small oil can, of the kind used for sewing machines, hold its spout directly over the circular hole in the paper fish, and let a drop of oil fall into it. The oil will immediately attempt to spread over the surface of the water, but, finding itself obstructed by the edges of the circle, it escapes through the narrow channel toward the tail, and is then released. As a result of the recoil effect, the paper fish, like a cannon that has just been fired, will follow a direction opposite to that of the oil flow, that is, it will swim forward. Try it for yourself; it will afford great pleasure to any children present, and to yourself as well!

The Camphor Boat

Our picture shows a small boat that has been cut with scissors from a piece of aluminum foil and then bent into shape. It is flattened at the back, or the stern, where there should also be a small cut in the shape of a wedge; if you try to copy the illustration, you will not find the task difficult in any way. To lend the end product a pleasing, eye-catching appearance, we can add a small mast made from a straw with a colorful flag flying from the top. Those who wish to go further may also people the deck with small paper sailors and a few brightly painted paper passengers, ladies and gentlemen from every corner of the globe, not forgetting to give the Captain his rightful position.

Our artistic preparations having advanced thus far, we can finally

turn to our experiment proper, which promises to be of great interest. We carefully place our little work of art in the water, and find that it sails splendidly. And if we then place a drop of alcohol into the wedge-shaped cut in the stern with a pipette or a pair of tweezers, as soon as the drop comes into contact with the water, the vessel will make a sudden move forward, so violent, in fact, that we have reason to fear our passengers may become seasick! The explanation of this effect – of the forward movement, that is, and not of the seasickness – is simple enough: at the bow and at both sides, the boat lies in unadulterated water, which offers far less resistance to its progress than the alcohol and water solution at the stern; it is this difference in pressure that causes the

boat to move forward.

The same pleasing effect can be obtained if instead of alcohol we use some other volatile substance, such as ether, chloroform, essential oils and so on. We saw this demonstrated earlier in the case of the swimming paper fish. The boat's motion will be more or less responsive depending upon the type of substance used. And if we place some camphor in the same spot, its vapors will not only produce the same effect, but also cause the boat to cruise about for hours.

FLOATING IRON

An interesting experiment is illustrated on the facing page and entitles even the layman to assert boldly that he can make iron float – in the stricter sense of the word, of course, for every child knows that iron ships have been floating across the seven seas for a good long time. But that does not concern us here.

To perform our experiment, we fill a wineglass nearly to the rim with water and ask the owner of the nearest sewing box if we may choose from the supply of needles one that is not too large – the sort that is found in the pincushions of industrious seamstresses the world over. Next we take a piece of writing paper from our desk and cut out a rectangle, one of whose sides is longer

than the needle. We place this piece of paper on the surface of the water and lay the needle on top of it. Now we shall be able to make the interesting observation that the paper sinks just as soon – and the time depends on its material properties – as it has absorbed a certain amount of water, in other words has become saturated. More surprising still to our audience is the fact that the needle will retain its floating position on the surface of the water even without its paper support.

The success of this experiment will be even more assured if we oil the needle slightly before we place it on top of the paper.

DRAWING A CIRCLE BY COHESION

IF we look at a glass filled with water, we tend to think that the liquid is evenly distributed throughout the container and that it is in perfect equilibrium. As we know, however, this is by no means the case. At least two forces are at work: attraction, which tends to keep the particles of the liquid together, and repulsion, which tends to separate them.

So far, we have been experimenting with fairly large quantities of water; let us now demonstrate cohesion with the help of a smaller quantity of liquid, but one that exhibits two lateral surfaces and hence can develop twice the normal

tension! To that end, we first dissolve one ounce of soap and one ounce of sugar in about three-quarters of a pint of water. Then we fashion a small square wire frame consisting of a simple loop on one side, wide enough to be grasped easily and surely between the thumb and index finger. If we now pour the solution into a dish, dip our frame into the liquid and lift it out again, we shall see that it is filled with a thin layer of water which seems to have hardly any weight, since it barely sags at all. The more liquid runs off, the thinner and smoother the layer becomes,

that is, cohesion and adhesion have counterbalanced the weight of the liquid, as you will no doubt have worked out for yourself. If we now tie the ends of an ordinary silk thread together and loop it deftly over this layer of liquid, the thread will retain its irregular shape and position; but as soon as we pierce the surface of the liquid in the center of the loop with the tip of a pencil or some similar instrument, the liquid will immediately withdraw to the edges of the loop and the thread will assume the shape of a perfect circle. A singularly interesting experiment!

THE SUSPENDED PENCIL

ADHESION only comes into play when bodies are in close contact with each other. It manifests itself between solids and also between solids and liquids. Thus two polished metal surfaces pressed together will adhere very firmly. This is the basis of the locksmith's well-known trick of grinding down two metal rulers so finely that, placed on top of each other, they stick as tightly together as if they were one.

Even more intimate is the contact, or rather the adhesive effect, if one of the bodies is a liquid. Thus, as the reader knows, if a solid is immersed in a liquid and pulled out again, a layer of liquid will generally adhere to it. In everyday life we say that the body has been made wet by the liquid. This effect, however, occurs only when the adhesion between the solid and the liquid is greater than the cohesion of the latter. Thus metal, wood, porcelain,

glass and similar materials are invariably made wet by water, wine, vinegar and many other liquids, but fatty or oily substances are not, as the reader well knows.

The following experiment is intended to demonstrate that, under favorable circumstances, the force of adhesion has a considerable carrying capacity. We take one thick and one thin lightweight wooden pencil, and place them lengthwise on top of each other, the thicker one uppermost. Now we pour a few drops of water into the gap between the two. On closer observation, we shall discover that the water has adopted the form of two small concave arcs *a* and *b*, as shown at the top of our illustration (page 104). At the same time, the adhesive force has come into play along both arcs with the result that the thinner, lower, pencil adheres firmly to the upper and thicker one.

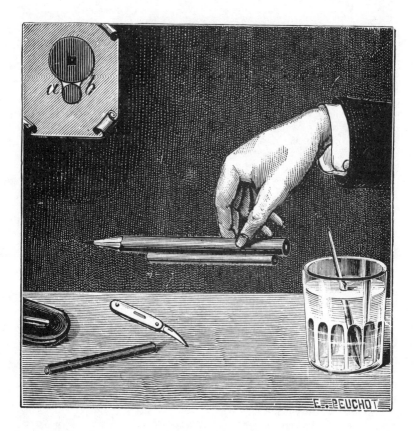

How to Fill a Sealed Wineglass

The reader may be familiar with the conjuring trick in which an empty wineglass is replaced by a full one without either of the two glasses being touched by hand, or in any other way. One of our quick-witted friends, who is always ready with a joke, used to deal with this problem simply by lifting the little table on which the two glasses stood, turning it around and putting it down again.

If the wine had also to disappear, then our "magician" would conjure up an obliging soul, who would throw himself upon the glass and empty it with obvious delight. The reader might be forgiven, therefore, for expecting our experiment to be based on a similar diversion; this is far from being the case, however, for we intend to stick to our more serious scientific games.

We take, then, two ordinary wine-glasses, with rims of equal diameter, and place one upside down on top of the other, as shown in the illustration, so that they appear to be hermemetically sealed. If we now pour a liquid, preferably water, very slowly from a third glass over the base of the upper glass, it might well be assumed that the liquid will run down the sides of the two glasses and end up on the table. To the astonishment of those not familiar with this experiment, however, this does not happen. Instead the water spreads over the base, trickles over the edge and down the sides of the reversed upper glass, and collects inside the lower glass. Before the experiment began we made sure that the two glasses were dried as thoroughly as possible.

This experiment demonstrates that the seal between the two glasses was not really perfect and that the water could find a way between the two rims, the displaced air inside escaping by the same route (*cf.* the experiment described on page 53).

A MUSSEL TURBINE

To make the handsome turbine shown in the illustration, we take a large round wooden box without a lid, which should not be difficult to obtain, and insert a small outlet pipe made of wood or metal near the bottom; it may be possible to find a discarded pipe stem which would do excellently for the purpose. To make the container, that is, the wooden box, watertight, the seams must be well sized and the inside of the base and walls painted with lacquer or oil paint. When this is done, we place a large button in the middle of the base; it should have a slight depression, as it will serve as the resting point or pivot for the verticle axle. Next we must construct two small vertical supports at opposite sides of the box and attach a strip of wood with a hole drilled in its center to their upper ends. This central hole must be exactly above the depression in the button on the base; it represents the upper support or pivot of the axle, which is slightly longer than the lateral supports and must be made in such a way that it can turn without hindrance at both top and bottom. We fit a thick and well-rounded cork, which as the illustration shows, will be taking up the driving belt, above the midpoint of the axle, that is, near the top. For the wheel itself, we take a larger round piece of cork or wood, cut out evenly sized teeth and attach mussel shells to them with small screws, so that some twelve such shells are arranged concentrically side by side. This wheel, too, is fitted onto the vertical shaft.

If we now wish to set our turbine in motion we need only direct a slender but powerful jet of water (either from a container placed higher up, or, better still, from a hose attached to the faucet) at the inside of one of the mussel shells. This appealing little machine – perhaps with the addition of another belt-driven machine – will then speedily start to operate.

A Paper Cornet Waterwheel

Our paper cornet waterwheel presents a striking spectacle, its primitive construction notwithstanding. It consists in the main of an octagonal wooden disk, which we must make first. To do so, we draw a circle with a pair of compasses on a small piece of board selected for the purpose. Inside this circle we draw a diagonal and cross it with another at right angles through the center. We divide the resulting four right angles with the pair of compasses in the familiar manner into two halves each, and draw a line through each dividing point to the circumference of the circle. Where these lines intersect the circumference, we draw in the tangents, eight straight lines, and our regular octagon is ready to be cut out with a fretsaw.

When we have made these preparations, we drill a hole in the center of the disk wide enough for an axle to be pushed through. Each end of the axle rests on a strong wire support fitted with an eyelet on top and pushed vertically into a fairly heavy chopping board. We now tack eight paper cornets of equal size, made from stiff but fairly thin cardboard, to the circumference of the wheel, making sure that they all point in the same direction. For greater durability, we can give the paper cornets and the wood a good coat of varnish before we start our little machine. The driving force is water, which we supply from a pitcher, having, if necessary, made suitable provision for the overflow. The little device can, of course, be powered by water from the faucet, and the wheel may also be used as an excavator by modifying the support and extending the axle. In that case, however, we shall need an alternative source of power, a longer axle for our wheel and the addition of a driving belt.

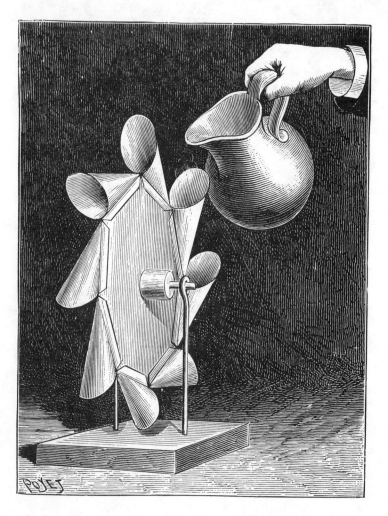

A Letter Balance for the Summer Vacation

We are taking our summer vacation in some mountain health resort. It rains and rains, which we naturally find most unwelcome, for we are hardly able to step outside the front door. So we write letters instead, in which we describe the glorious summer weather at length, in the hope that the envy of those left at home will compensate us somewhat for our spoiled vacation. This is an arduous undertaking, and the letters grow fairly long and have to be weighed.

Because there is no letter balance available, we have no option but to make one for ourselves. We have enough weights, for we know that a U.S. cent weighs exactly $\frac{1}{10}$ ounce. For the balance itself we use an old broomstick found under a seat in the summerhouse; quickly we saw off a piece of the handle about 10 to 12 inches long, tack a visiting card to the top end and attach a small heavy metal object to the lower end with a piece of tape and a small nail. Then we obtain a tall narrow vessel, fill it with water and immerse the stick in it. Since our weight is not very large, the piece of wood with its cardboard "pan" will stick out of the container. Now we place ten one cent pieces (exactly one ounce) on the visiting card, and watch the stick sink deeper into the water. With a pencil we mark the level of the water, and check it again by taking the coins off the pan and then replacing them. Now we remove the coins once more,

place the letter on the pan – and find that the stick does not sink down as far as the pencil mark. No

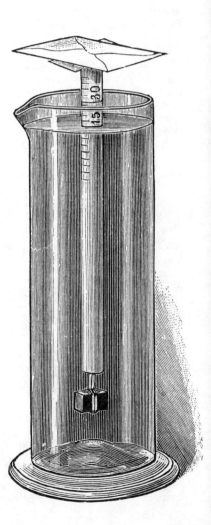

extra stamps are needed, and – hey presto! – the letter is in the mailbox. And so our dear ones at home are saved the trouble of having to reduce the deficit of the post office by paying postage due.

THE SNAIL-SHELL WATERWHEEL

WE first procure the shells of twelve to eighteen edible snails, then use a fretsaw to cut out two plywood disks of equal size and push a dowel through their center to form an axle. We attach the snail shells to the circumference of both disks so that they all point outward and in the same direction; this, we readily admit, is quite a delicate operation. Wire and sealing wax will not do the job properly, and we have to resort to a different bonding agent, perhaps a good all-purpose glue, being careful, of course, not to shift the shells while we glue them in place. We must also insert small rectangular wedges between the axle and the two wheels, gluing them to both so that the disks do not revolve like wagon wheels on their axles, but are firmly attached. Now we place the axle of our waterwheel on the prongs of two uprights attached to the base and provide a gutter at the bottom. We then push a large cork over the projecting end of the axle, across which we run the driving belt

which is best made of stout canvas of the appropriate width.

Now a few brief words on the operation of the waterwheel: if a jet of water from a large vessel with a long, narrow spout at its base is directed at the openings of the snail shells, the latter fill with water and move downward under its pressure to empty into the gutter and the container placed underneath. If it is intended to run the machine continuously, perhaps to keep a second little workshop busily occupied, then a constant flow of water must be provided.

A HYDRAULIC MOTOR

HERE we have a toy that can be made with a minimum of effort out of the simplest material but that can nevertheless be put to practical use. All we need for its construction is four walnuts, a wooden board with four posts at the corners, several rods, a fairly large cork, reeds and tapes of various lengths, and sundry containers.

Let us begin with the four posts, which must be fixed very firmly to the board because they will be supporting the entire mechanism. In particular, they have to accommodate two square, fairly thick driving shafts. The front shaft will carry the wheel and must be slightly stronger than the transmission shaft at the rear. The wheel consists of a central body, six spokes and as many scoops. The body is made from a large cork, which must be as solid and nonporous as possible for maximum stability. The six spokes are attached to it at equal intervals and in the same plane so that any pair of opposite spokes forms a straight line. To the ends of the spokes we attached the nut shells. This must all be carried out with great precision and everything fitted together very firmly so that our "motor" can run smoothly. Apart from this wheel, we also attach a smaller transmission wheel, again made of cork and, facing it, on the second shaft, another transmission wheel, and connect the two wheels by a fairly taut band or belt. To the front end of the transmission shaft, we fit the device illustrated on page 112; but more about this later.

The power is provided by a water container and a siphon, both of which must be placed higher than the rest of the equipment. The higher the container, the more power will be generated. The siphon, as our picture shows, is made from a walnut and three lengths of reed. The spout must, of course, be twice as long as the suction tubes. So that the siphon is kept firmly in position, the suction tubes should be attached to the side of the container or else weighed down. If we suck water through the spout, it will continue to flow as long as the suction tubes remain immersed in the liquid. Once the siphon has been activated, we position it in such a way that the stream of water falls onto the scoops of the waterwheel. The overflow is caught in a container placed beneath the wheel;

as soon as it is full, it can be used to replenish the upper reservoir.

As the reader will see from the illustration, the artist has combined the apparatus with a little butter churn. To make the latter, we need a suitable lidded container for holding the cream. Next, we construct the device that we mentioned earlier. This consists of a disk with a number of perforations, which moves in two directions, vertically and horizontally, as it beats the cream. The range of both movements is exactly double the length of that of the peg attached to the shaft; the strokes can therefore be lengthened or shortened at will. In the experiment illustrated here, however, it is best not to choose too long a stroke since otherwise the vessel containing the cream would have to be too wide, a factor which would not be favorable to butter.

EXPERIMENTS WITH CAPILLARITY

As the reader well knows capillarity results from both the mutual attraction or cohesion of the particles of one and the same substance, and also from the mutual attraction, or adhesion of the molecules of different substances. It is because of cohesion that a drop of water or mercury retains the shape of a small sphere when placed on a greasy plate, instead of dispersing, and it is because of adhesion that we need force to separate two highly polished glass or metal plates in close contact with each other. The joint effect of both these forces, that is, of adhesion and cohesion, thus explains the phenomenon of capillarity, at least in part. The same effects are responsible for a whole series of striking phenomena, some of which are described below.

Experiment 1: Fold the sides of a very thin piece of paper to make a flat box with a depth of about an inch. If a small amount of water is poured into this container, after it has first been moistened on the inside with a wet brush, the water will curve inwards at the edges, causing the sides of the container to become bowed.

Experiment 2: If we make a soap bubble with a funnel and then release it, it will shrink perceptibly, because the molecules of the liquid tend to move closer together. The effect becomes even more obvious if we stop blowing before the bubble has finished forming, and leave the bubble suspended on the funnel: the bubble will flatten out and even withdraw inside the funnel.

Experiment 3: We hang one thin wooden rod from another by threads and suspend them from a small hook, so that the rectangle $A B C D$ (Fig. 3) is formed. If we now dip the whole into the solution we used earlier for making the soap bubbles, we shall find that the shape of our figure has been altered: the contracting force of the soap skin has curved the threads on either side of the rectangle inward and has hence drawn the rods closer together (Fig. 4). The tendency of the liquid to decrease its surface area can thus be plainly demonstrated.

A Simple Prism

If we allow sunlight to pass through a prism, we may observe the following effects: 1. the light rays are deflected (refracted); 2. the composed (white) sunlight is split into its spectral components, each of which is deflected differently, red least of all and violet most; 3. the spectrum which consists of seven main colors (red, orange, yellow, green, blue, indigo and violet), takes up more space than the cross section of the original white beam; 4. the boundaries of the spectral colors differ from the corresponding boundaries of the white light: they are curved, whereas the former were straight.

These effects are usually studied with the help of polished glass prisms. We can, however, substitute a cheap water prism for the expensive glass instrument by allowing the rays of the sun to fall perpendicu-

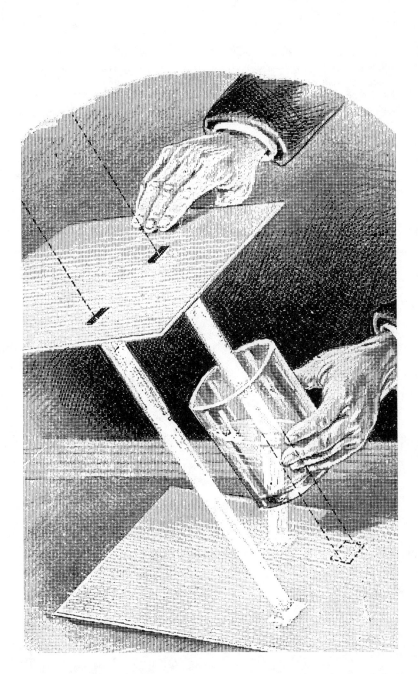

larly on a piece of cardboard into which we have cut two rectangular windows, obtaining as a result two parallel beams that produce two rectangular white images on the white paper lying on the table. Now we take an ordinary cylindrical tumbler with a plain base, fill it one-third full with water and hold it at such an angle beneath one of the cardboard windows that the beam from the window constitutes the axis of the cylinder. Since the surface of the water is no longer parallel to the surface of the base we have produced a splendid water prism. We may study all the effects we have described with one of the two beams while the other will help us to determine the degree of the deflection and of the color dispersion.

SUNDIALS

ON the assumption that the reader is familiar with the principles governing the construction of sundials, we shall briefly mention two that may be produced by very simple means.

The illustration below depicts a strip of paper that has been varnished or soaked in oil, divided into hours, and glued to a cylindrical glass tumbler stopped

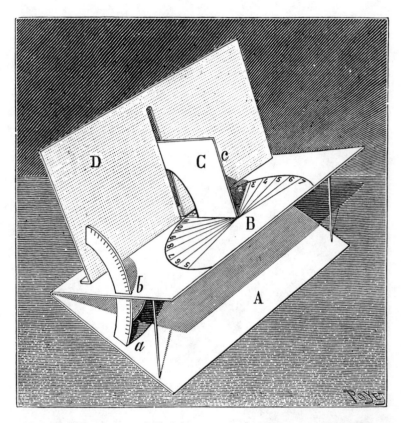

with a piece of cork or wood (*b*). A metal rod (*c*) is inserted vertically in the middle of this disk to form the pointer whose shadow will indicate the time in accordance with the position of the sun. The tumbler rests on a hinger board *E F*, which can be set at a slant as required with the help of two small props.

Of similar construction, but more practical, is the sundial depicted on this page, which can easily be constructed from ordinary cardboard. It consists of two pieces of cardboard, *A* and *B*, glued on the reverse side

to a piece of canvas, with a dial drawn on *B*. There is a folding top *D*, with a central slit through which the pointer *C* may be pushed and attached to *B*. We insert the protractor (*b*) through a slit in one of the shorter sides of *B*, attaching it to *A* beneath; its purpose is to position *B* in accordance with the local meridian, the props (*a*) serving to maintain this position. The pointer *C* when lit by the sun casts a shadow on the dial from which the time can then be told.

THE REFLECTION OF LIGHT

LIGHT is known to physicists as matter consisting of extraordinarily small particles. It is assumed that these minute particles, though weightless, are subject to the general law of intertia. Their dimensions and relations with other bodies have been determined, together with a series of effects involving their powers of attraction and repulsion; these

investigations, however, are by no means concluded; many luminous phenomena have not been fully explained to this day.

The reader will be familiar with one of the most common effects in this sphere, namely refraction, thanks to which a stick immersed in water will appear bent at the point of immersion. This happens because the light rays which the stick transmits to the eye are deflected as they leave the water.

If light travels from a denser to a rarer medium, we may observe what is known as total internal reflection. It is thanks to this effect that air bubbles in water look like shining pearls, and it also explains the brilliance of many precious stones: part of the incident beam passes through the stone and reemerges at so sharp an angle that it becomes totally reflected and the stone assumes an almost metallic sheen as a result. Of all precious stones, the diamond is known to have the smallest critical angle, that is, the smallest angle of incidence beyond which light undergoes total internal reflection, and this explains its exceptional brilliance.

The following is an easily performed experiment: if we look at the flame of a candle through a glass of water, with the eye slightly lower than the water surface, then we shall behold the reverse image of the candle against a dark background. This is due to the reflection of the slanting rays, which having hit the surface of the water cannot continue in a straight line.

THE REVERSE IMAGE

EVERY one of our readers probably knows that a camera obscura produces reverse images, that is, turns objects upside down. The same is true of all the images received by our retina, our eye, in fact, acting as just such a camera obscura, except that our brain corrects the mistake. We mention this fact at the outset since it forms the basis of the explanation of the experiment we shall now describe.

Take a visiting card and pierce a hole in it with a pin. Hold the card some four inches away from your eye and place a pin between the card and the eye. The image of the pin will disappear from in front of the card but reappear, standing on its head, to the rear of the small hole we have made (see the illustration, top right on page 120). The pin must be viewed against a window or a lamp. The small hole serves merely as a slit for admitting the light (Fig. 2A, this page) which casts the shadow of the pin upside down onto the retina.

We can achieve the same result in

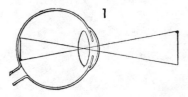

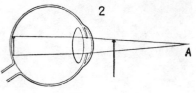

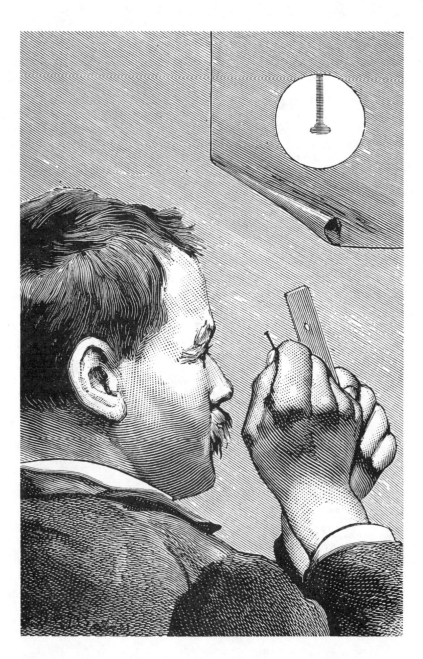

a different way. If we hold the hole in the card very close up to our eye (about $1\frac{1}{2}$ inches away) and almost close the eye, we shall see our eyelashes upside down, that is, pointing upward. And if we move the card to and fro in front of the eye, or rotate it, a network of little branches will appear after a while in our field of vision. What we have "before our very eyes" is the image of the network of veins in the retina. (This particular experiment does not always succeed at first try.)

Right- or Left-Sighted?

"Whatever can be meant by 'right- or left-sighted?' " many of our readers will wonder. It is well known that most people are more agile with the right hand than they are with the left, but that with some the opposite is the case. Left-handed people are feared with good reason on the fencing ground because they are unnatural opponents for the ordinary right-hander. But most people also believe that what is true of the hands does not apply to the eyes. After all, we generally use both eyes, whereas we do not always work with both hands simultaneously. Indeed, it is generally supposed that the positioning of our eyes gives us a great advantage over most animals, since we are able to view the same object from two different view-points, and as a result to identify its physical shape, properties and importance at a single glance, while a bird, a chicken, for instance, may have its suspicions about an object but is unable to make up its mind until it has examined the object with first one eye and then the other, repeatedly turning its head to do so.

In general it is true that we rely on the impressions gained visually by both eyes, much as we lift up very heavy objects with both hands. However, if need be, we can also see with one eye; indeed, there are occasions on which it is impossible to see with both, because the visual impression of the one interferes with that of the other. Whoever looks through a monocular telescope or a microscope, whoever sights along a rifle while hunting or in the army, will need only one eye since there is no work for the other to do. But the layman very often finds himself in difficulty in such cases, being unable to stop himself from using the other eye! The most obvious thing to do, of course, is to shut it, but many people can only do that by holding one hand in front of it, or by pressing the eyelid down. Or else, having taken note that those practiced in this art can close their other eye without using their hand, they try vainly to follow their example, pulling all sorts of faces in the process. Now it is a moot point whether the eye that is not involved should be shut, or whether it can remain open. If both eyes are open, a picture is naturally formed on each retina; but do we necessarily have to take cognizance of both images? Often enough, we find ourselves staring into space while our thoughts are busy with something quite far removed; we do not give our surroundings a thought, that is,

we take no notice of our visual impressions, so that afterward we have no idea at all of what our eyes have seen. That this "non-seeing" can also affect just one eye when both are wide open is shown by the simple experiment depicted in our first illustration. We make two tubes from cardboard or paper, about 1 foot long and some 2 inches in diameter. Next we place two different sheets of newspaper on the table and hold the tubes in front of our eyes in such a way that we can look at one sheet with one eye and at the other sheet with the other eye. We shall not be able to read both pages at once, of course, but can focus when we like on one or the other, although both eyes remain open all the time. It takes a simple act of will for one eye to see properly and for the other to stare blankly.

It may thus be argued that we often make use of only one eye, even when we keep the other wide

open, and, accordingly, that we do not have to exclude the other eye by holding a hand in front of it or by shutting the eyelid, even when aiming a gun, looking through a microscope or engaging in any other activity requiring the use of only one eye. However, it has been shown that we do not use either eye at random for seeing, but that we always prefer one to another. The human race can be divided into right- and left-sighted people.

But how can we decide whether a particular person is right- or left-sighted? This, too, can be settled by a simple experiment, which we shall once again illustrate with a picture. Pierce a hole in a sheet of paper with a pencil, place a second sheet of paper on the table some 18 inches away from the eyes, and draw a black circle (about the size of a small coin) on it. With both eyes wide open, hold the first sheet of paper between your face and the table, slightly nearer the latter, and move it about until the black circle can be seen through the hole. Next, keeping the paper still, close first the left eye and then the right. In most cases, the circle will disappear when the right eye is closed, indicating that we could see the circle with both eyes or with only the right eye open, but not with the left eye alone. In other words, when both eyes are open, the left does no work, so that the viewer may be said to be right-sighted. If the reverse takes place, then we have a left-sighted person.

We know that there are many more right- than left-sighted people. We may conclude, after some reflection, that there is some casual connection between this phenomenon and the right-handedness of most people, though we cannot really say which is the cause and which the effect. Since time immemorial, man has drawn a bow, pulled a trigger and flung a stone with his right hand, using the left for support or, in conjunction with the arm, for keeping his balance. If we watch a boy aiming a stone at a target, we shall see that he lifts the stone level with his eyes, inclines his head slightly to the right, follows through the straight line from his eye to the target with his right hand and finally throws the stone at this level with his right arm. This is, no doubt, how it has always been, and it is probable that hereditary factors have played their part in moulding this preference for the right eye.

It might be thought that the unequally sharp vision of the two eyes implies a preferential use of the keener one. This is undoubtedly so over long distances when those who are short-sighted in the right eye will invariably use the better, left, eye. But this does not mean that they are left-sighted, for as soon as they look at an object nearby, the right eye comes into action and takes over from the left. Needless to say, those who are blind in the right eye are necessarily left-sighted. Such people will point a gun on their left or, if they have to point it on the right, will turn their heads to aim with the left eye. Any left-sighted person will have to do the same, or else close the left eye firmly, thus forcing the right to sight the target.

A Trick with Light

In a darkened room, we move a table up against a white wall or against a plain curtain; on the table we place the cover of a note-book in which we have previously cut two stars. Both stars are four-pointed and congruent, although one is turned 45° in respect of the other. In front of them we place two candles, the tops of whose wicks are level with the central points of the stars. We must experiment with the positions of the candles on the table in order to ensure the coincidence of the centers the stars cast on the wall. If we temporarily screen both lighted candles, and then remove one of the screens, a four-pointed star will

appear on the wall. If we then remove the second screen as well, the star will be transformed into an eight-pointed one. We shall notice that its – doubly illuminated – center is particularly bright and, what is more, in the shape of an eight-pointed star, although one that does not correspond to the large star either in shape or in position: the points are shorter and blunter and lie between the points of the larger star.

This phenomenon, which bears a distant resemblance to the fade-over effect in photography, may be modified in various ways. Thus we can magnify the light of one of the candles, and hence the brightness of one of the four-pointed stars, by placing a globe-shaped glass filled with water in front of it. Or else we can hold a red slide in front of one candle and a blue one in front of the other, whereupon the star will have four red and four blue points, and a violet center. Resourceful readers will no doubt be able to work out further variations of their own; what we have tried to do is to encourage you with the help of an easily performed experiment.

SIMULATING SPECTRAL COLORS

FIG. 1 on the right depicts a disk which should have a diameter of exactly $1\frac{1}{2}$ inches; one half of the disk must be completely black, and the other half must have four concentric sets of curves, three black arcs to each set. The disk must be placed in quick rotation, for which purpose we can use the rotation device depicted in Fig. 2, top, or a drill. At a given rate of rotation, which differs from one observer to the next, the brightly lit disk will display concentric rings of color, although the disk itself has nothing but black-and-white components. This effect seems inexplicable at first sight; but it soon becomes clear that what we have here is a natural phenomenon: as everynone knows, white is composed of all the colors of the spectrum, and these colors can be separated, as we have learnt from the prism. What we have before us, therefore, are artificially produced spectral colors. But why do the individual constituents of white light manifest themselves in this particular experiment? We shall try to suggest the reason.

In Fig. 1, let L be the lens of our eye, and let the straight lines impinging on it from the left be rays of white light. Now, as the reader knows, the individual components of white light are refracted in different ways, red the least, violet the most. Let us confine ourselves to these two constituents, and if we slightly exaggerate the dimensions in our figure, we shall

find that the red constituents merge in R, the violet in V. The retina is situated at E, between R and V. In order to see a red object against a black background, which is only possible if the image falls on the retina, we must adapt our eye in such a way that its lens becomes more curved, so that, as a result of increased refraction, the junction R of the red rays is shifted toward E. By contrast, to see a violet object clearly, we must relax the lens. A white object against a black background cannot, strictly speaking, be seen clearly or sharply, for if the lens is strongly curved we obtain a clear red image R with a faint violet edge v, and if it is more relaxed, the image becomes violet (V) inside a red edge r, and if we imagine a median adaptation (accommodation) of the lens then neither red nor violet will predominate; instead the image (b) will be blurred around the edges and too large.

Now let us rotate the disk in an anticlockwise direction, say, five times a second. During the first one-tenth of a second we shall see the black semicircle, for which no accommodation is needed; then black-and-white stripes will appear alternately at the edge of the disk, the white stripes with a red rim merging into black. Now accommodation comes into play, but slightly overshoots the mark. As a consequence, the more central stripes appearing after one-fortieth of a second will look white with a yellow edge merging into black. After a

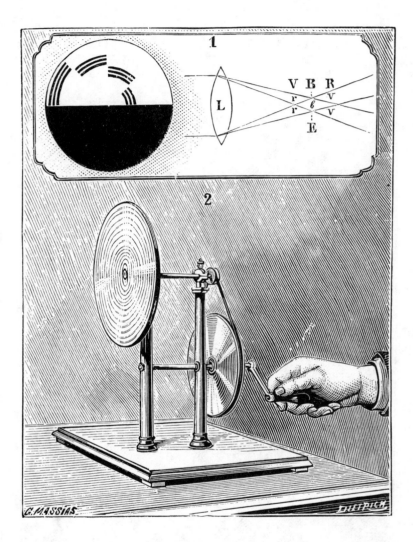

1

V B R
r   v
r   v
E

2

G. MASSIAS   DIETRICH

further one-fortieth of a second, the
third group appears in the form of
white stripes with green edges,
followed by a fourth group in the
form of white stripes with blue to
violet edges. Because of the quick

rotation, however, the eye sees full
circles, not just small arcs, of a
particular color. The distinctness of
the effect will depend on the powers
of accommodation of the eye, which
varies from one observer to the next.

It may frequently happen that we want to magnify a small insect we have caught, or the opalescent wing of a butterfly, or some other object, and if we have no lens or microscope to hand, or perhaps do not own them at all, we are in some difficulty.

Our illustration and the accompanying description are intended to demonstrate that in these circumstances the reader has no need to abandon the project, provided he is in possession of a glass globe. With its help he can fashion a lens, admittedly very primitive but serviceable all the same, and thus obviate the need to acquire expensive equipment.

We take our glass globe, which if possible should be joined to a glass tube, fill it with pure water and plug the mouth of the tube tightly enough to prevent the water from leaking out when we turn the tube upside down. Next, as is shown in the illustration, we take a length of wire, winding one end several times around the tube, and bending the other upward so that it is level with the center of the globe. If necessary, we can file that end to a sharp point for spiking the object we want to examine, or a piece of paper on which we have previously mounted it. We have solved our problem, for if we now hold the glass globe in front of our eyes like a magnifying glass, we shall find that the insect or specimen on the other side is considerably enlarged.

A Distorted Drawing

In a spherical mirror, everything seems reduced in size, and the further the object is from the mirror the smaller is the image. This optical phenomenon is associated with the convexity of the reflecting surface. Since a sphere is curved evenly all over, the objects it reflects appear evenly reduced in size, so that the images look like true miniatures of the objects. Things are quite different in a convex mirror that is not perfectly spherical, for instance in a cylindrical mirror. In cross section, the latter will show the same spherical curvature as the sphere, but its longitudinal section is a plane surface. As a result, a cylindrical mirror acts lengthwise like a plane mirror, that is, it reflects everything in its real size, while sideways it behaves like a spherical mirror.

When we look at ourselves in a cylindrical mirror, we appear our natural size, but as thin as a rake, and if we turn the mirror round so that the top is at the side, we appear to be dwarfs but preserve our natural girth. Both effects are distorted images that make us laugh, and that is the reason why such mirrors are found in so-called fun houses in amusement parks. Oddly enough, we consider the distorted image in the first case as too thin, and in the second case as too broad rather than too short.

Let us go back to the first case: the mirror does not alter our height at all, but decreases our width. If we now draw a human figure which is intentionally much too broad in relation to its height, and hold it in front of the mirror, the latter will

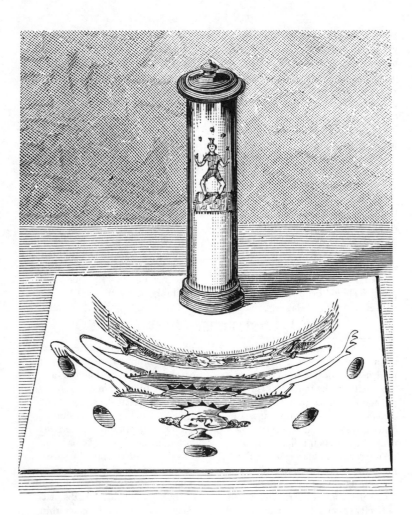

cure the corpulence of the drawing very simply by decreasing its width in the way we have described, perhaps by exactly the right amount to arrive at a normal human body. In other words, the mirror has converted a distorted image to normal proportions. Things are not so easy if, instead of holding the drawing in front of the mirror, we place it underneath. The height is then somewhat affected as well, and curvatures of various kinds are produced. If all this can be compensated for in the drawing, and the mirror is placed on top of it, we see a perfectly normal, distorted mirror image of the drawing.

THE MAGIC MIRROR

GHOSTLY apparitions on most illusionist stages are based on the use of a pane of glass, which, like all glass plate, has the double property of reflecting and transmitting light. The old platinum magic mirror is based on the same principle: it consists of a plate of glass coated on the back with a layer of platinum so thin that the glass remains transparent. It is possible to produce patterned glass with reflecting inscriptions, decorations or figures in the same way. For this purpose, the glass is first coated with platinum over its entire surface, and then the platinum layer is inscribed with the desired patterns or letters (in mirror writing), and the drawing etched in.

Of particular interest, too, are the ordinary magic mirrors mentioned at the beginning, which are variously transparent or opaque, depending on whether they are lit from the front or the rear. A large pane of glass is fitted into an ordinary mirror frame which has had a large opening cut in the back and is fitted with a door that will open and shut. This magic mirror must not, of course, be suspended from the ceiling but must give onto a niche or some other recess so that the door at the back of the frame can be opened and the glass illuminated from the rear. When this is done, it will be possible to see through the glass everything that was previously hidden behind the mirror. Thus, if a large picture or similar object is placed behind the pane, the viewer will not notice it while the frame door is closed; all

he will see is his own mirror image. But if the door is suddenly opened and very bright (perhaps electric) light is shone onto the glass plate from the rear, the viewer's own mirror image suddenly vanishes and in its place there appears whatever has been placed in the empty space behind the frame – our picture shows a devil's head. His surprise will certainly be quite comical!

There are sundry other experiments along these lines. For instance, if a magic mirror is fitted into the frame of a door whose panels have been removed, we shall have a glass door with a very curious property: if it divides a brightly lit room from a dark passage or antechamber, the latter will be illuminated from the room although, to anyone inside that room, the glass will look like a mirror. We can also coat the inside of spectacle lenses with platinum, so that the wearer is able to see through the lenses which, from the outside, look like opaque mirrors. In the same way, we can replace window panes with one-way mirrors that allow us to look out into the street, but nobody to look inside.

No doubt a great many other uses for the magic mirror may be devised, the most difficult aspect being the procedure by which the platinum is applied to the glass. For this we procure some concentrated platinum chloride and sprinkle oil of lavender over it. We then paint the resulting viscous fluid onto the glass with a good brush, taking great care not to let any dust settle. Next, the

platinum has to be baked on. The mirror is placed in a drying room for a short while, transferred to a baking oven, and finally allowed to cool slowly. If it is then rubbed with a linen cloth and a little permanent white, it is ready to be put to the many uses we have mentioned.

MIRROR IMAGES WITHOUT MIRRORS

As everybody knows, not only mirrors produce reflections, but any sheet of glass placed against a fairly dark background will reflect rays from a brightly lit object, and hence make it appear as a more or less clear mirror image. We can easily prove this to ourselves by strolling past the display windows of a store.

The experiment depicted in our illustration provides similar proof. If we place a pane of glass in a vertical or near vertical position on a table in a darkened room, and put one playing card in front of it and another behind it, adding a candle for better illumination, then after a few trials and errors we shall find exactly where to place the cards so that, looking at them from the front, we can see them both side by side, one directly, through the glass, and the other beside it as a mirror image.

This simple fact can help us to practice extremely convincing deceptions, and many a traveling showman has known how to exploit it to full advantage. Thus we may be sitting in the audience at a fairground booth and fail to notice, as the curtain rises, that the poorly lit stage is separated from the even darker auditorium by a huge pane of plate glass inclined at an angle of about 20°. We watch the scene unfolding behind the glass. Then suddenly, in the course of some bloodcurdling action, a ghostly figure looms up and glides over the stage toward the hero, who, showing incredible courage, runs his sword through the figure or fires a gun at it – all without in any way endangering the life of someone who, in a coordination with the hero, has created the surprising effect from a sunken area in front of the stage, effectively deceiving the audience.

THE CABINET OF MIRRORS

IN our search for interesting scientific curiosities, we discovered a device that we would specially commend to all those anxious to combine physical with amusing effects. What we have in mind is not so much a novelty as a rearrangement and magnification of the old kaleidoscope. The apparatus is extremely simple.

We need three large unframed mirrors of equal size, which we place vertically in such a position that their bases form an equilateral

triangle and the mirrors constitute a three-sided prism with their reflecting surfaces facing inward. We must also arrange some access to our mirror cabinet, either from below, from above or else by moving one of the mirrors aside. In the arrangement depicted in our woodcut, entrance was made through a trapdoor in the floor.

When anyone steps inside this mirror cabinet, he will see his image being reflected over and over again. This effect is due to the fact that, in a mirror, the angle of reflection of a beam of light is equal to its angle of incidence, so that, with our present arrangement, the image of an object is reflected as an infinite number of symmetrically arranged groups whose points of incidence and reflection lie in the corresponding extensions of the planes of the three mirrors. Our illustration shows no more than a small part of the mirror images cast by three visitors inside the cabinet. These three people saw their figures reproduced in infinite sets of six persons each, the reflections growing increasingly smaller until they appeared to be lost in the dim horizon. Needless to say, the effect will be even more vivid if six people step into the cabinet.

HEXAGONAL CIRCLES

"WOULD you think it possible that the little white circles on the black background you can see in the picture could be made to look hexagonal?"

"No, I wouldn't!"

"You are a doubting Thomas, but you'll soon learn."

We set up the page so that our doubter is looking at the circles from a distance of three to six feet.

"Well, are the circles still round?" we ask him.

Reluctantly, he will have to admit that they now seem to be so many hexagons; indeed, if he is a careful observer he will also inform us that the spaces between the circles are no longer black, but gray or even lighter.

This experiment is based on the so-called irradiation effect, thanks to which the edges of the circles seem to overlap and hence the circles appear larger and to have lost their true shape.

A related optical illusion can be produced with a white triangle on a black field and a black triangle on a white field. You may pride yourself on your eyesight, but all those who are asked which of the two triangles is the larger will decide in favor of the white one, although the two triangles are in fact congruent.

PENUMBRAL PICTURES

WHEN parents and children gather around the family table on long autumn and winter evenings, the little ones in particular invariably enjoy a show of shadow pictures. All that is needed for conjuring up the drollest of figures in this way is a bright light, a white wall and a pair of nimble hands.

Those who are more advanced in this art may like to add paper silhouettes, and since the toy industry has not as yet, or only just, discovered shadow play, we must make them ourselves. This form of self-expression can be recommended for its own sake since, as the saying goes, if you want a thing done well, do it yourself.

We feel that we will earn your gratitude if we enlarge your reper--toire with the following little known, though not new, act. When light rays are allowed to fall on a paper disk, a dark shadow or umbra surrounded by a penumbra is produced behind the paper, and even if the silhouettist is not an expert, he can use this phenomenon to produce the most striking effects. Let him take a piece of stiff card and, using our first figure as a guide, draw a picture. He must then cut out the parts that are to look bright, and if he now holds the remainder close to the wall on which the shadow is cast, the umbra will predominate and the silhouette will have a sharp, clean, outline (Fig. 2). But as the cutout is moved further and further away from the

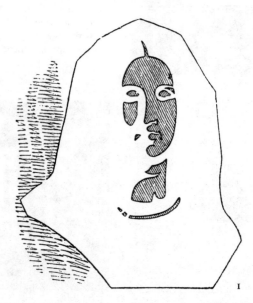

I

wall or screen, softer shades will be added to the umbra (Fig. 3), rounding the portrait off in a surprising and most pleasing manner. By careful calculation of the distance and meticulous cutting of the card it is thus possible to produce shadow pictures of considerable artistic merit.

2

3

The Devil in Reverse

So-called positive afterimages may be simply and easily perceived by anyone looking fixedly for a short time at a brightly lit object, the frosted bulb of an electric lamp, for instance, and then closing his eyes or looking at a dark background. The resulting afterimage is brightest of all if we view the object for no more than one-third of a second. In addition, there are negative afterimages, which reproduce the bright sections of the object as dark areas, or conversely, reproduce the dark sections as bright. We can obtain these if we stare at a bright object continuously and then, instead of closing our eyes, direct our gaze at a white or otherwise bright surface. The parts of the retina that were previously exposed to the dark areas of the object retain their full sensi-

tivity. As a result the contrast
between bright and dark is completely reversed.

An amusing example of this
effect is afforded by our illustration:
a white imp on a black background
that may be easily reproduced
should this book not be close at
hand even by those with a minimum
of artistic skill. If we stare at this
white inhabitant of the nether
regions, concentrating particularly
on the black spot in the middle
until our eyes tire, and then gaze at
a bright surface, say, a sheet of
paper, or, better still, a white ceiling,
a rectangular white surface will loom
up before our eyes within which a
black devil will gradually emerge.
The experimenter should not be
discouraged if the devil does not appear immediately; some people may
even see the devil before they see the
surface. In any case the reader
would do well not to put too much
strain on his eyes, that irreplaceable
organ of vision.

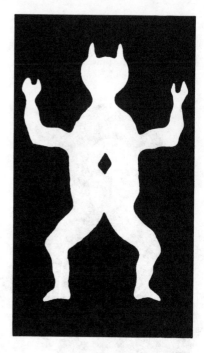

THE FADE-OVER EFFECT

THE term "irradiation" refers to the
fact that the light rays emanating
from particularly bright objects not
only excite those parts of the retina
on which the image is formed, but
also affect the neighboring parts of
the retina as if the latter had received
the light themselves. Hence the
impression in the brain is such as to
make the object look larger than it
really is.

To prove this point, the reader
need only view the black and white
ladies in our illustration from a
distance of about five feet; he will

find that the black lady on the white
background looks more slender and
petite than the white lady on the
black background. Indeed, the
illusion is maintained to some extent
even when the ladies are viewed
from close quarters. And yet both
ladies are exactly the same size, as
measurement will easily establish.
Were our illustration life-size, the
effect would, of course, be more
striking still.

If we have a cloth with alternating
black and white stripes of the same
width, the black stripes will appear

narrower than the white. This can be
demonstrated if the figure on the
right is magnified by a factor of ten.
If this figure is viewed from a
distance of ten to twelve feet, the
white stripe will look much broader,
and also longer, than the black.

Finally, we should like to mention
an irradiation phenomenon that may
be reproduced with great ease.
Following the illustration on p. 142,
draw a narrow black cross on a
square of white paper. If the square
is observed very intently, the ends of
the cross seem to retract.

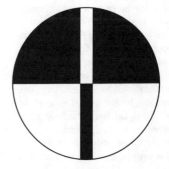

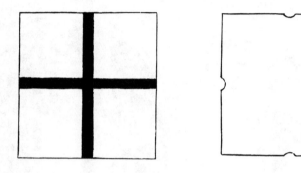

How Tall Is the Top Hat?

The entertaining experiment illustrated on page 143 always arouses a good deal of surprise. One of those present is asked to estimate the height of the demonstrator's top hat by pointing to the spot on the wall he thinks the hat would reach if it were set down on the floor.

At first our volunteer will look at us pityingly, as if to say: "What do you take me for? My eyes aren't nearly as bad as you seem to think."

"Let's put it to the test, then. Off you go!"

Our friend will shrug scornfully, take one more look and indicate a spot on the wall. We take the hat off and hold it on the floor to check the accuracy of his judgment.

Great merriment all around!

Our friend, however, is somewhat embarrassed for, as everyone present can see, he has overestimated the height of the hat by at least a handsbreadth. And all others who are not in the know will without doubt fare the same, some even indicating a spot twice the height of the hat.

In fact, all of us estimate the height of an object differently, depending on whether we view it at eye level, or at an angle. The reader will, of course, be aware that from the top of a tower or monument people appear very much smaller than when they are seen at street level.

We can try another amusing experiment with our top hat, by asking those present how tall they think the hat is in relationship to its width. They will be particularly cautious following the embarrassing experience of the first volunteer, having learned that the laugh is always on the loser. Each in turn will voice a different opinion, one guessing that the hat is $1\frac{1}{2}$ times as tall as it is broad, another saying $1\frac{1}{3}$ times, a third $1\frac{1}{4}$ times, and so on. We can be fairly sure that no one will come up with the right answer, namely that the height of the hat is the same as its width, unless someone in the

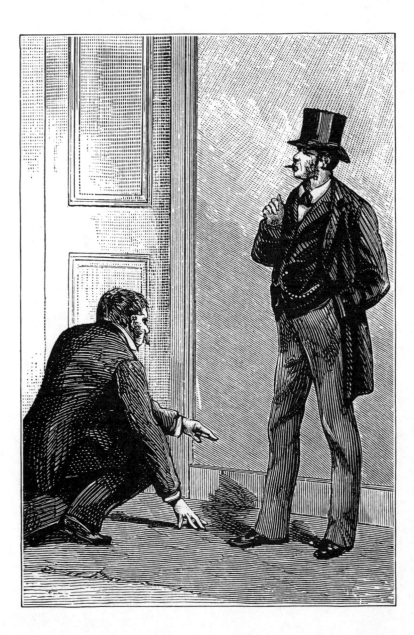

audience happens to be a hatter, who will have good reason for being in the know.

But why this widespread misunderstanding?

It is simply that people do not usually take the turned-up brim into account, for if they did they would quickly realize that the width *AB* is equal to the height *CD*, as may easily be checked by measuring the dimensions of the hat depicted on page 144.

OPTICAL ILLUSIONS WITH THE HELP OF A RULER

As we saw earlier, optical illusions can be created through the use of the so-called irradiation or fade-over effect. The latter is based on the fact that illuminated surfaces not only seem larger than dark surfaces of the same size, but that they may also look quite other than they really are. If we hold a flat wooden ruler up close in front of our eye and against the flame of a candle or a lamp, as the keen young man is doing in our illustration (p. 146), then the ruler will seem to have a concave indentation on the line running from the eye to the flame. In other words, the light of the flame seems to encroach upon the ruler at this spot, with the result that the latter looks "waisted."

The same effect may be observed if we are in a darkened room and look at a slit or small aperture that admits a beam of light. It will seem that the aperture is much larger than it really is. The same effect will also appear if we look at the ridge of a roof against the setting sun; the ridge will seem curved even if it is perfectly straight. In these last two examples, the light encroaches upon the dark, or not directly illuminated, parts of the objects in question and apparently alters their appearance. This fact is well known to painters, sculptors and civil engineers.

THE MAN IN THE BOTTLE

THAT our illustration is a faithful woodcut reproduction of an actual photograph may be hard for you, dear reader, to believe at first. We can assure you, however, that the process to which we owe this surprising and original oddity is extremely simple; the following directions will enable every amateur photographer to produce similar effects, and with some degree of ingenuity and skill he may even succeed in surpassing our own achievement.

To photograph the man in the bottle, we place the latter, which, needless to say, must be transparent, in front of and fairly close to the lens. Some distance away, beyond it, we stand the man against a dark background and on a platform covered with a black cloth.

We then make sure, by looking through the viewfinder of our camera, that both objects appear to be the same size, that is, that the bottle seems to enclose the man, moving the man or bottle back-

ward or forward if necessary. We now photograph man and bottle separately and successively on one and the same negative. The film is developed in the usual way, and the positive developed.

If the man should happen not exactly to despise a brimming bottle, and to be, moreover, a friend who will not take a bit of fun amiss, then we will use, not a medicine bottle, but a clear white wine bottle, and, if we append some humorous verses to the photograph, we shall have a welcome and amusing family or birthday keepsake.

THE GIANT HARE

WHEREVER did these two lucky fellows obtain their gigantic hare? The photograph on which our illustration is based was apparently obtained in a single shot, so that our friend the hare must have been of quite extraordinary size – a miracle of nature that makes our naturalist friends shake their heads incredulously. And they are right to do so, for this photograph, too, is based on a trick.

It is well known that hands crossed in front of the chest or stretched out toward the camera look disproportionately large on photographs, the reason being that they were closer to the lens than the rest of the body. Our snapshot is based on the same principle. For when it was taken, the hare was not hanging from the stick carried by our two intrepid hunters, but from a fine wire much closer to the lens of the camera than the men, and so placed that the hind legs of the hare appeared to cross over the stick when examined in the viewfinder. As a result, friend hare gained in size over the hunters, as the photograph shows, and a single shot was enough to produce this result. But what about the equipment from which the hare was suspended and which cannot be seen in the picture? The animal was simply hung from a dark wire that ran down the middle of the dark tree trunk in the background and hence remained quite invisible in the photograph.

Amateur photographers among our readers should not find it too difficult to copy this example, should an opportunity arise, or to crack similar photographic jokes for their own enjoyment and that of others.

THE CANNIBAL

HERE is another joke photograph, which shows a guest who, with the imperturbable composure of a cannibal, has just picked up his knife in order – or so it would seem – to start feasting on the head that has been served up to him – his own!

The answer to the puzzle is the same as that described in the previous experiment, and is far from difficult to achieve. The photograph is taken against a dark background, provided here by a doorway giving onto a dark storeroom. The lens is focused on the background, and the head of the man, who is standing fairly close to the camera, is observed through the ground-glass viewer. It's relatively large outlines are then drawn on the ground-glass plate and a blackened piece of cardboard is placed in the first fold of the bellows (counting from the ground-glass plate) so that the rest of the body is blocked out. Next the plate is inserted and a photograph of the head is taken.

The plate and the sheet of cardboard are now removed, and preparations are made for the rest of the picture.

The man sits down at the table further away from the camera than before, and the table and platter are positioned in such a way that the head, drawn in outline on the ground-glass plate with a pencil, appears to be lying on the platter. The plate is then reinserted and a second photograph is taken. As a result, the head will seem to be lying on the dish, just as shown in our illustration.

Your Head on a Platter

THE most curious effects can be produced by composite photographs, that is, by making two exposures on one and the same plate. For this and the following experiments we first need a plate-back camera with a bellows, and then a dark background, say an open door giving onto a dark room. This was the basis of the photograph below, in which a guest is offered his own head on a platter, to his ill-concealed horror.

To produce this kind of effect we must proceed as follows: after focusing the lens on the dark doorway, we ask the "guest" to move close to the camera so that his head will look disproportionately large. Next we take a pencil and draw the outline of his head on the ground-glass plate, cut a hole the same size and shape in a blackened sheet of cardboard and place the resulting stencil in the first fold of the bellows (that is, right in front of the photographic plate) so that only the cut-out area (the head) is lit up. Then we introduce the photographic plate and

take a picture of the head. That plate is now removed, and the rest of the photograph prepared. The guest takes his seat with an expression and attitude of horror, and the servant holds a napkin-covered platter in such a way that the head drawn on the ground-glass plate appears to be lying on it. (We have, of course, previously taken the cardboard out of the bellows.) We put the plate back in again and take another photograph. The picture of the head is not affected in any way because of the dark background, and the end result is our very strange picture.

Once again we leave it to those of our readers who are fond of experimenting to devise other compositions – perhaps a dinner party where all the participants are the same person.

PHOTOGRAPHIC COMPOSITIONS

ALTHOUGH photographic collages can nowadays be created much more easily with the help of ingenious darkroom techniques, the reader may be interested to learn something about the older method from which the wood engravings shown on these on the right.

While the composite pictures of the preceding pages made use of dark backgrounds, the following pictures, taken against various backgrounds, were created with what is known as a "frame with adjustable slit." Let us first examine this surprisingly simple invention (on the right).

It consists in the main of a tightly fitting wooden case attached to the back of the camera with an aperture the size of the plate at its center. This case supports a long frame protruding above and below the camera (Fig. 1, *E*). Inside this frame and beneath two short boards, there is a longer, black board that can be moved from the top of the frame to the bottom, and that is provided with a central slit. Fig. 2 shows the outside and the back of the

frame. If a photographic plate is inserted in the camera, and the slit is moved up or down past the plate by the steady rotation of the crank, the plate will be exposed wherever light falls on it through the slit. It is essential to turn the crank very steadily so that the plate is evenly exposed. It should also be noted that the board with the slit has a pointer

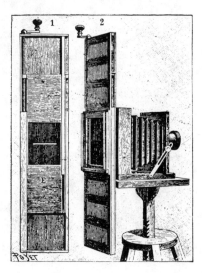

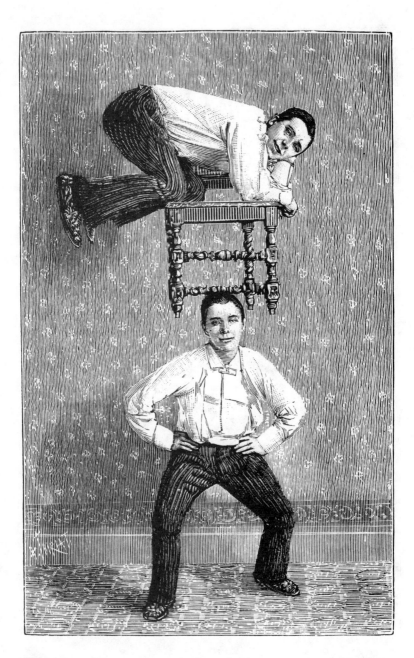

and a graduated scale (front edge of frame, Fig. 2) indicating the position of the slit at any given moment.

It is a very simple matter to produce composite photographs with the aid of this adjustable device. If we look at the two gymnasts, one of whom carries the other on his head, we shall see that they are one and the same person. The picture is the result of two exposures. To create it, the chair is fastened to the wall at a height of four to five feet, and the gymnast climbs onto it, taking up the required position. The photographer examines the image, section by section, through the slit while turning the crank, and notes how far the pointer must move along the scale until the seat of the chair and the tops of the chair legs come into view; in addition he draws the outlines of the chair on the ground-glass viewer. Next, he inserts the photographic plate, and turns the crank once again from the starting position until the pointer has traveled as far as before. At this point the camera is shut, the plate is removed, the gymnast climbs down from the chair and takes up his position underneath the chair. The photographer compares the position of the chair with the outlines drawn on the viewer to make sure that the chair did not move when the gymnast climbed down. The plate is reinserted, and the second part of the photograph is taken, the crank being turned so that the pointer runs down the rest of the scale.

The photograph on page 153 is much more difficult to take. It consists of four poses by one and the same person, and involves three (instead of two) horizontal zones, the central one being divided into a left and a right half, each of which must be taken separately. To that end, the slit must be provided with a slide capable of shutting off its left or right side temporarily, or else the whole frame must be turned through 90°, so that the slit becomes vertical and can be moved from left to right, or vice versa.

SHADOW PICTURES

BEFORE we examine true shadow pictures, we should like to mention a technique that enables even the poorest of draftsmen to catch the likenesses of his friends with telling verisimilitude. We take a sheet of paper, white on one side and black on the other, and pin it to the wall so that the white side faces the room. Next we place a bright light on a table at an appropriate distance from the wall, and ask the person whose portrait we are about to draw to step between the light and the wall and to stand perfectly still. Because our subject will cast a sharp silhouette on the paper, anyone can easily move a pencil around the outline of the shadow. We can then remove the paper from the wall, go over the lines again where they may not be clear, and cut out the drawing. Now all we have to do is to reverse the cutout, stick it onto a white piece of paper, and the result is a black silhouette. If we have deployed even the minimum of skill, the likeness is generally a very good one.

After some practice, even the beginner will acquire a measure of skill in this technique and produce portraits as good as the ones shown at the top of our woodcut. Moreover, those who can handle a pantograph will be able to reduce the size of these silhouettes at will, and thus perhaps produce an album to treasure for many years to come.

Let us now turn our attention to shadow pictures proper, for which we need nothing more than a white wall, our two hands and a bright light. Our fingers must not, of course, be stiff and awkward, or else we shall produce only the most unshapely monstrosities. Should any readers wish to devote themselves to this delightful and amusing art, and not at times possess the requisite agility, they must practice, in the knowledge that, with persistence, they are bound to be rewarded with success.

Let us begin our performance with the reproduction of a fox. The positioning of the fingers is extremly simple, as our illustration shows. The fox's muzzle, long and pointed, reflects cunning and slyness, and is

slightly agape with expectancy. Since no prey is available the fox opens its jaws wide and yawns several times, but we need have no fear that our attentive audience will find that infectious and do likewise (below).

Following this demonstration of our skill, we portray a human silhouette, namely an oarsman guiding his craft safely through ripple and wave while traveling gently downstream. The right hand is easily bent into the shape representing the man; a small piece of wood attached to the thumb

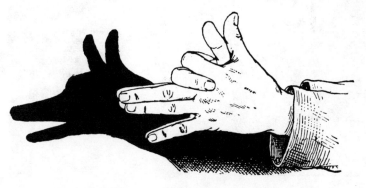

represents the oar, a paper cutout held between index and middle fingers, the hat.

Now comes Mr. Hercules. This strongman holds a weighty dumbbell in his hands and a chair between his teeth. The doll's house will supply the latter; the dumbbell is made from a small rod and bread pellets.

The elephant is a most comforting and friendly animal. Its eye is particularly appealing, it swings its long trunk to and fro like a pendulum, rolling it up and then pointing it straight out in front again. Those who wish to add to the effect can enlist a third hand to feed it an apple (a bread pellet); it will grasp the apple with its trunk and devour it with obvious enjoyment.

And here comes the cheerful tippler. We use the index finger of

the right hand to form the bulbous nose of the habitual drinker. The little finger casts the shadow of the strongly jutting chin on the wall, while the hat, once again cut from paper, is held between thumb and index finger. The stretched-out arm with the bottle is formed by the left

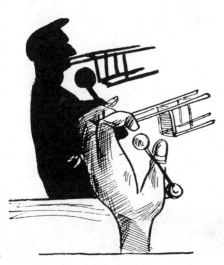

hand and part of the arm; a piece of cloth falling from the right to the left arm makes the body. Needless to say, the tippler is allowed to indulge his passion, taking a swig or two now and again.

Now for the juggler, who performs a most astonishing balancing feat with an egg on a stick. For the stick we need a long match; the egg is cut out of stiff paper and fits into a slit in the match made with a knife. The hat, also cut out of paper, is tucked between index and middle finger, and the match together with the egg is tied to the thumb.

The charming rabbit needs the participation of both hands, as the illustration below makes clear. This graceful animal is extremely lively: it moves its ears and drums vigorously with its forelegs; it can even sit on its hindlegs and beg.

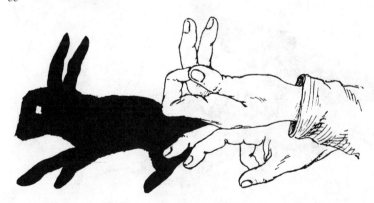

The North American Indian, too, must not be left out of our repertoire. The leading role here is played by the right hand, which has to shape the face; the left hand is bent with fingers spread to produce the silhouette of the feather finery.

For the bull we need both hands. The projecting muzzle is formed by the right hand, while the left represents the head and the threatening horns.

The German naval officer is slightly more difficult. The right hand has the main job, the formation of the face, its ring and little fingers forming the walrus moustache. The helmet is made with the left hand, thumb held vertically upward and little finger protruding.

The wicked wolf has, it is true, screwed its eyes shut, but it has a

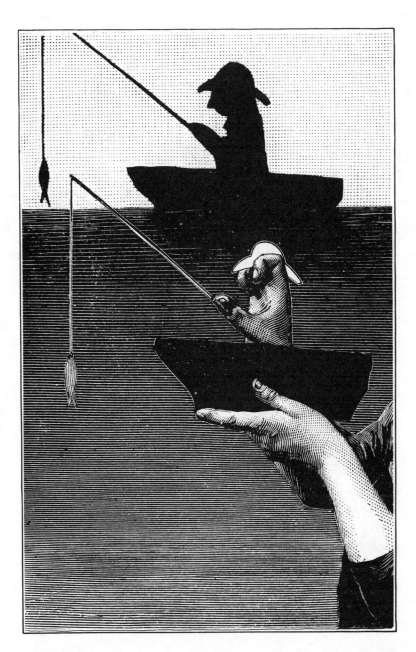

terrifying set of fangs. The latter, as the illustration shows, is represented by the finger tips of the right hand. When the wolf opens its jaws wide and snaps them shut again, it truly makes one shiver with fear and dread.

The angler, by contrast, has a much more peaceful look about him. He is part of a scene in which practiced fingers deftly reproduce the bobbing boat, the eager anticipation of a good catch, and finally the exultant reaction to the sight of a fish on the end of the line – all of which will delight the spectator (see the picture on page 161).

The Punch and Judy show (page 162) which does call for more than average skill, should also evoke loud applause from the audience.

A great deal of merriment will be caused by all these shadow pictures, as we can assure the reader from our own experience, and this effect will be enhanced even further if the performer is able to link the pictures with an amusing commentary. He will, however, have to acquire, by constant practice, a certain amount of skill in the quick formation of successive shadow pictures before he can hope to command the attention of a happy circle of spectators.

The four silhouettes of well-known late nineteenth-century political figures (pages 163, 164, 165) are intended, by way of a conclusion, as encouragement to you, dear reader, to try your hand at reproducing the silhouettes of contemporary statesmen.

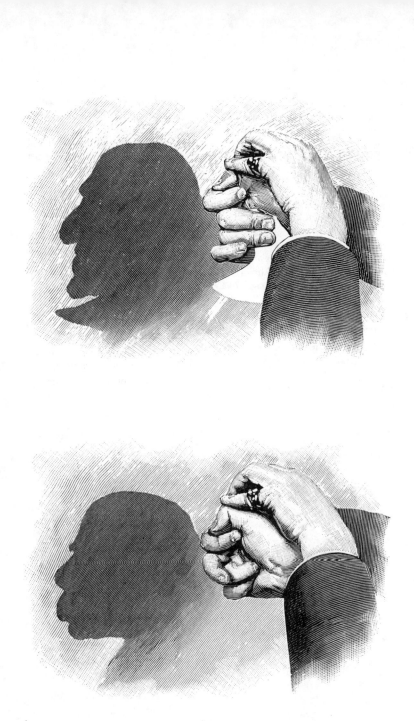

164

The Singing Doll

The droll experiment depicted in our illustration is admirably suited to evoke great merriment at the family table, particularly amongst the younger members, although adults, too, are sure to be entertained. To the knuckles of the index and ring fingers of either the right or left hand, we apply a few bold strokes of charcoal to draw two eyes and two nostrils, as shown in our illustration. The thumb, pressed tightly against the index finger or turned slightly downward by it, serves very well as a toothless mouth. The knuckle of the index finger forms the nose, above which the two eyes are located. If this face is now wrapped in a cloth, it will take on a close resemblance to the physiognomy of a toothless old woman, especially in a dim light. The artist can practice moving the thumb, which forms the lower lip and the chin, in time to a song (warbled in the nasal manner of old folk) or a piece of dialogue, so that the old woman's mouth appears to open and close, and, at intervals, she can cough or sneeze with the appropriate accompanying movement. No matter how serious the mood of the company, this act is bound to cause laughter.

The illusion becomes complete if the artist himself stands behind a suitably hung linen cloth, so that he cannot be seen by the audience, and holds the doll above this curtain, as in a Punch and Judy show.

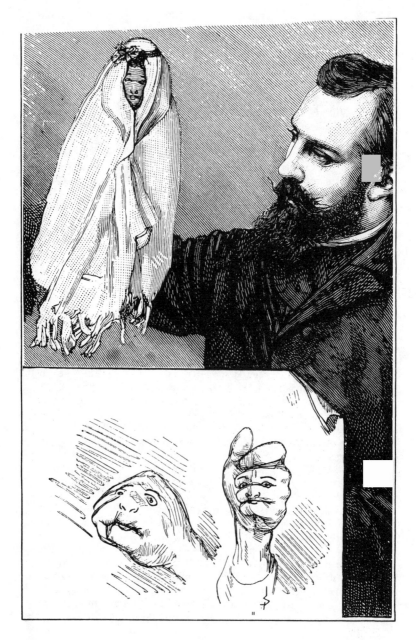

Playing-Card into Chain

Is it possible for a small piece of card, like an ordinary playing card, to be cut in such a way as to produce a chain more than three feet in length, and for the links made in this way to be joined without any glue? In fact, anyone in possession of a penknife with a very sharp blade and a pair of scissors can perform this trick, provided he takes care. He should examine the figures below closely and then set to work:

1. Split the card along the edges A and B with the penknife to a depth of about $\frac{1}{4}$ inch; this is easily done, since every playing card consists of two sheets stuck together;

2. Bend back the open edges A and B so that a perfectly straight crease is produced;

3. Fold the card across the middle from C to D;

4. Cut the doubled-over card with the scissors along the vertical lines marked in the illustration, at intervals of a $\frac{1}{4}$ inch as far as the split and the turned-back edge.

5. Open out the card, fix it to a drawing board and insert the penknife alternately over and under the cut-out strips, at the same time lifting them from one of the split edges.

6. Proceed on the other side in the

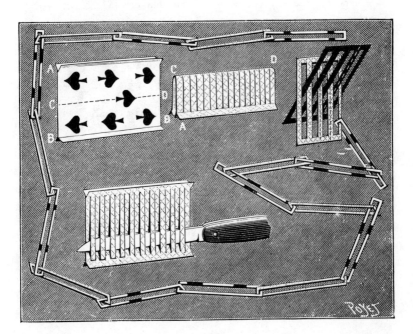

same way in reverse, so that alternate strips are cut free from the edge. Now the card can be separated into two intertwined and linked parts.

7. Finally, cut one link after another lengthwise with the scissors along the lines shown to obtain a closed chain of eight, ten or twelve links, depending on whether the card was cut into ten, twelve or fourteen strips.

A Trick with String

TAKING a piece of string some six feet long, and very pliable so that it may easily be bent into all sorts of loops and coils, we hold the two ends between the thumb and index finger of the right hand and make the figure that is shown on the left of our illustration. Those who find this too complicated can arrange the string on the table before picking up the two ends (*A* and *B*). Now we set ourselves the task of pulling the string off the table while a volunteer attempts to stop us by placing an index finger at any point in the string figure that he wishes. The chances are a hundred to one that our volunteer will choose that part of the figure marked with *o*. When we pull the ends of the string, he very quickly realizes that the string slides smoothly past the tip of the finger that is meant to hold it back.

Let us start afresh!

We arrange the string on the table again, and now say that if the

volunteer places his finger once more on the same spot he chose before, his finger this time will be caught and he will have succeeded in what he originally set out to do. The volunteer will again place his finger on the spot *o*, convinced that we are about to be proved wrong. We pull, and his finger is well and truly caught in the string.

How did this come about?

The volunteer fails to notice that we have made a slight change in the way we arrange the loops of string on the table; it is only when he looks at the two figures side by side and compares them carefully that he will notice the difference. In the figure on the left, the right string leads straight into the loop *a*, whereas in the figure on the right it leads first into the next loop, so that loop *a* now appertains to the left end of the string.

THE VANISHING BALL

LET us now turn our attention to black magic, which, although it is not quite as "black" as it seems, still calls for some practice and skill from the performer.

The magician produces a small, multicolored ball, which he allows the spectators to examine closely, and then places a transparent glass box on the table, puts the ball down beside it, covers each with a different cloth and promises that, upon his

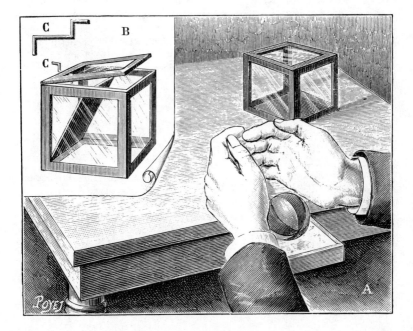

command, the ball will travel into the box. After tapping the latter with his magic wand, he pulls the cloth over it away, and displays the ball in the box. Now he pulls away the second cloth and shows that there is nothing underneath it. The magician again covers the box with a cloth, while he ostentatiously crumples up the other, taps the box with his wand, pulls the cloth away and the ball has disappeared. When the crumpled cloth is lifted up, it is seen that the ball is not there either. The magician looks astonished, goes up to one of the spectators and pulls the ball out of his pocket.

The magician has, in fact, been using two balls of the same size and color, one solidly stuffed with horse-hair and the other elastic and yielding to the slightest pressure; it is the first that he displays to the audience. The cube-shaped box (B) is framed in wood, fitted with hinges on the top for opening the lid and provided with six glass walls; opposite the lid, that is, on the floor,

at a height of about $\frac{1}{2}$ inch, a polished nickel plate is hinged to one of the sides and, when lifted against the vertical wall, can be held in position with the clasp C, acting as a catch, so that the rubber ball lying between the plate and wall cannot be seen and the box appears to be empty.

The solid ball, after having been shown to the audience, is not in fact covered with the crumpled cloth, but discreetly dropped into an open drawer (A) at the back of the table. The box is then covered with the other cloth, the catch is turned so that the metal plate releases the squashed ball which then resumes its spherical shape. During its second disappearance, the box is turned while the cloth is placed over it so that the rubber ball is under the metal plate on the floor of the box and hence cannot be seen, while the other ball, which the magician has previously concealed in some part of his clothing, is skillfully pulled out of the spectator's pocket.

INSEPARABLE COMPANIONS

THE magician asks one of his audience to provide him with a small key and a handkerchief. He loads the key into a blunderbuss, places a small box tied up with string some distance away, fires at it and hands it to the owner of the key for safe keeping. He then pushes the hand-kerchief into a bottle, but quickly pulls it out again with the excuse that he has chosen the wrong bottle.

The handkerchief is now a red color. The magician apologizes, and, placing the handkerchief on a dish to dry, sets fire to it by mistake; the handkerchief is completely burned. He expresses his deep regrets, and asks how much it cost; he then loads the charred fragments into the blunderbuss, and fires them at the box. He now asks for the latter to be handed to him, cuts the string

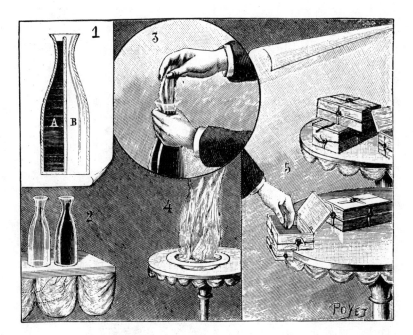

around it, and pulls another tied-up box from inside it, that box holding yet another, and so on. The last box contains the key and the undamaged handkerchief.

The blunderbuss is an ordinary stage pistol, with the bell of a small tin trumpet fastened to its mouth. It is loaded with a blank, and, while this is being done, the key is dropped into a bag or basket attached to the far edge of the table. The bottle (Fig. 1) is partitioned into two equal sections, *A* and *B*, of which *B* has no floor; *A* is filled with spirit dyed red with aniline, and a thin piece of cloth inserted well before the performance. The borrowed handkerchief, carefully rolled up, is pushed into section *B*,

the bottle with the open bottom is pushed over a bag fastened to the edge of the table (Fig. 2) which already contains the key, and the magic wand is pushed down to force the handkerchief into the bag, whereupon the dyed cloth is pulled out of the bottle (Fig. 3) and burned in the dish (Fig. 4). The last two of the nesting boxes have no floor, and the smallest is kept hidden at the edge of the table. While pulling out one box after another, the magician pushes the penultimate discreetly over the last, in which he has just as discreetly secreted the handkerchief and the key, cuts the strings and pulls out the two objects to the not inconsiderable astonishment of all those present.

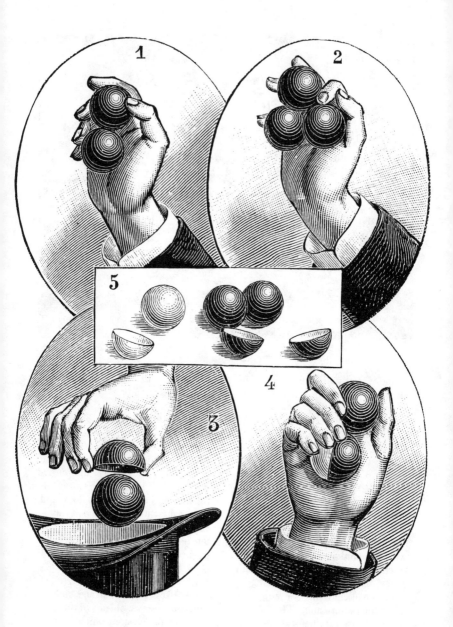

Magic Spheres

The three small balls are made of wood varnished in various colors, and go with three thin tin hemispheres that fit over them tightly and must be colored to match (Fig. 5).

While the magician displays one of the balls, rolling it across the table, he picks up the appropriate hemisphere unnoticed with the other hand, orders the ball to stop, quickly fits it into the hemisphere, picks up the two together, holds them both between thumb and middle finger, and then lets go of the ball, making it look as if he had been holding two balls in his hand (Fig. 1). While he is bent over, ostensibly to pick up the ball, he fits it quickly back into the hemisphere, picks up another ball with his free hand, and nimbly slipping off the hemisphere, displays three balls to the astonished audience (Fig. 2).

The magician now announces, holding two balls in his hand, that one of them will appear mysteriously inside a hat. He points to the latter, and nimbly drops the ball from the hemisphere into it (Fig. 3); then, still showing two balls in his hand to the audience, he puts the hat to one side, and quickly fits the hemisphere in his hand over the sphere, displaying the second sphere in the hat to the audience. Similarly, he can pretend to drop one sphere through the hat, by throwing in the hemisphere and at the same time dropping the ball onto the floor. While bending down to pick it up, he quickly removes the hemisphere from the hat.

The magician then displays one red and one white ball, with the hemisphere on the white ball so that the audience is looking at two red balls. By nimble sleight of hand he now produces balls of matching and different colors alternately (Fig. 4), finally dropping the hemisphere unobtrusively into a basket fastened under the rear edge of the table, whereupon he hands the balls around to show that they are solid.

The Mysterious Knot

"Ladies and gentlemen! You see before you two identical ribbons. Would one of the ladies present be kind enough to lend me her bracelets for a short time? She will have them back quite undamaged. I now thread the ribbons through the bracelets, and ask you, Sir [one of the spectators], to hold these ends of the ribbons in your hand, and you, Sir [another spectator], to hold the other ends. Now, do you think that it will be possible for me to free the bracelets while you are holding the ends of the ribbons, indeed even after the ribbons have been knotted around the bracelets for greater safety? Please be good enough to hand me one end of the ribbon, say the right one, and you, Sir, the other end, that is, the left one. I shall tie a knot and return the ends of your

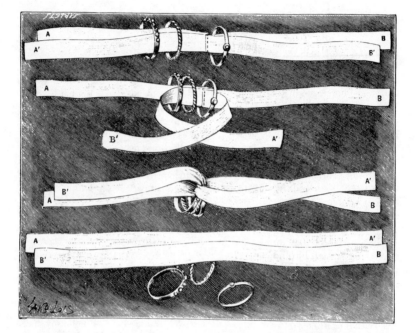

ribbons to you. Please don't pull too tightly. I now touch the knot with the tips of my fingers and, hey presto! the bracelets drop off, the knot has disappeared and the ribbons are undamaged."

For this trick, we need two ribbons of the same color, length and thickness. We fold one (AA') end to end, and give it a sharp crease in the middle with a hot iron. We treat the second ribbon (BB') in the same way. We then place the two creases on top of each other so that the ends A and A' lie on the left and the ends B and B' on the right. Next, we sew the creases together with a thread of exactly the same color as the ribbons. The thread may have a small knot at one end but, for the rest, the seam must be loose so that the thread

may be easily pulled out by the knot. We now draw the ribbons through the bangles (see the top illustration), and hand the ends to our two spectators. Each will be holding both ends of a single ribbon, rather than the ends of two ribbons. The tying of the knot leave the spectator, who previously held A and A', with A and B', and the other spectator, who had held B and B', with B and A'. This mix-up is sure to go unnoticed, but as a result the bracelets with the knot around them are now held together by the seam alone. If we finally, while apparently brushing the knot with the tips of our fingers, pull the thread out of the ribbons, the bracelets will immediately drop off.

THE VANISHING BIRDCAGE

THE magician shows the audience a birdcage and then picks up a small box into which he places a few small objects he borrows from members of the audience, such as a ring, a coin, or a postage stamp, supposedly for use later as a sign that the birdcage has not been changed over. The effect is greatly enhanced if it is possible to procure a ring large enough to slip around the neck of the bird inside. The cage is then covered with a large cloth which is waved around and finally grasped at two corners and tossed up into the air. Both cage and bird have gone! The magician continues for the time being with another trick, but sud-denly seems to remember the bird-cage, listens out for a few seconds, and then claims that he can hear the bird sing, suggesting some spot, for instance the room next door, in which the cage will be found. Some-one goes to look, and finds that the magician has been telling the truth. The magician now takes the little box out of the cage and returns the borrowed objects to their owners.

For this trick we need a small birdcage about 8 inches long, 6 inches broad and 5 inches high, a large enough cloth that has been doubled over and sewn at the c corners to a wire frame *A* (see the illustration). This frame must be the

same size as the base of the cage. When the cloth is displayed to the audience, it is held up by the two corners of one edge in such a way that the shape of the frame does not show through. Behind the table, which is covered with a cloth and placed fairly close to a door, there is a large upholstered chair onto which the magician slides the cage when he lifts the cloth.

Now the magician steps in front of the table and distracts the attention of the audience with another trick while an assistant unobtrusively takes the cage through the curtain-covered doorway and places it wherever the magician has decided that it will eventually be found. Since the cage itself has no secret devices, it can be handed to the audience for closer inspection.

THE MAGIC CHAIN

It is often advisable, during an enthralling performance of tricks based on sleight of hand, to introduce a short entr'acte so as to divert the spectators' attention. The so-called magic chain illustrated for the reader on the next page is particularly well suited for this purpose. It consists of a large number of intertwined rings, and since the general effect is based on an optical illusion the trick may be performed without much practice.

Once the chain has been displayed to the audience, the magician picks up the top ring A (Fig. 1) with one hand and pulls the ring B to one side with the other, whereupon the chain will automatically fall into the position shown on Fig. 2. And if ring A is then suddenly released, it seems to the observer that it slides down all the way from the top of the chain to the bottom. What actually

happens, however, is that ring C, and all the other rings in turn below it, twist over one by one, each dropping down one link. Since this all happens very quickly, the audience thinks it has seen ring A traveling down the length of the entire chain, and is deeply impressed by what it considers an amazing feat.

This type of chain can be made by any plumber, indeed by any one of our readers, except for the soldering, which is best left to an expert. The rings are of tin-plated iron or brass wire, which is wound tightly around a rolling pin twenty or thirty times and then cut lengthwise along a straight line. The resulting rings are arranged and linked as shown in Fig. 2, and the open ends soldered together. It does not matter whether the number of the links is odd or even.

THE COIN IN THE BOTTLE

Most people will be astonished when a magician produces a bottle containing a coin so large that it could not possibly have been intro-

duced through the narrow neck.

After this amazing bottle has been passed from hand to hand, the magician asks if anyone feels able to

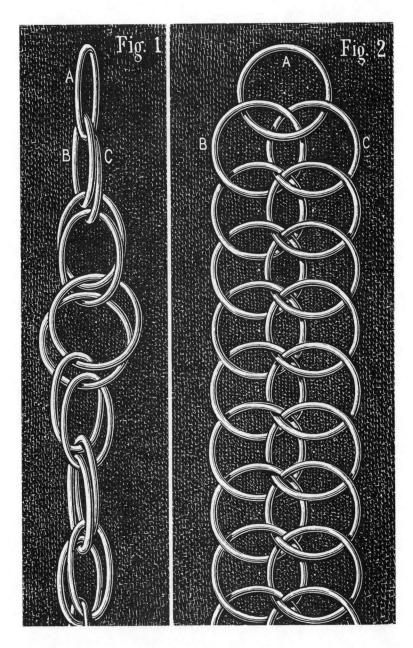

extract the coin, for which purpose an ordinary small pair of pincers is provided. The audience answers with a shake of their heads, although some smart aleck will probably offer the suggestion that the bottle be smashed releasing the coin that way.

The magician now steps up to the table with the bottle, turns the bottle upside down, reaches inside the neck with the pincers, and the coin drops onto the table. He hands around the coin and the empty bottle as proof of his achievement and is bound to earn general

acclamation from his astounded audience.

The solution is very simple: the experiment calls for two identical coins whose diameter must not be greater than two and a half times that of the inside of the neck of the bottle. The magician cuts three grooves with a file on the edge of one of the coins, and then uses a fretsaw to divide the coin into three almost equal parts, cutting carefully around the outlines of the head, eagle or other device, so that each part can easily go through the neck of the bottle. While sawing, special care must be taken to follow the profile and the outline of the back of the head as closely as possible, so that the cuts remain indiscernible. The three parts are now fitted back together, heated to a fairly high temperature and held closely in position with a tight rubber band, which will adhere after cooling. As the "coin" is pushed into the bottle, one part bends forward and another bends back, but all will return to the right position once they are inside the bottle thanks to the pull of the rubber band. The coin is removed in the same way with the pincers, and quickly exchanged for the other kept in readiness, which is then handed around amongst the audience.

WHERE IS THE DIME?

THE magician spreads a freshly laundered, colored handkerchief on the table, asks for a dime and places it in the middle of the cloth in full view of the audience. He then folds each corner of the handkerchief in turn across the coin, and as an extra precaution asks one of the spectators to feel the spot so as to make sure that the coin is still in place.

"Hocus-pocus abracadabra sintoque fix!" These words flow solemnly from the magician's lips, in the tone usually employed by magicians at public performances. The first corner of the handkerchief

is pulled away, then the second, the third, and the fourth. General astonishment – the dime has disappeared.

How was this done?

The performance of this piece of magic is simplicity itself; the explanation can be found in our illustration. Near the corner of the handkerchief he first lifts up is a dark spot, on which the magician must press a small piece of wax, kept hidden in his hand until the right moment. As soon as the wax sticks to the material, it is pressed onto the coin in the center of the handkerchief, and the other corners are folded over it. When the corners are eventually unfolded, no one will notice that the magician does not remove, as he ought, the fourth or topmost corner first, but the first corner instead, to which the wax and coin now adhere, quickly secreting the coin itself in the palm of his hand. He then very slowly lifts the second, third and fourth corners, and to the astonishment of the spectators, the coin – "Hocus-pocus abracadabra sintoque fix!" – has disappeared.

THE BEWITCHED SPHERE

A WELL-KNOWN French magician used to cause a great sensation whenever he performed with a large metal sphere which he persuaded to move up and down in time with the music on a piece of string wound around his fingers, and which he now and then brought to a sudden halt. The way in which this effect was produced used to puzzle his audience a great deal, and his light wooden sphere has ever since been a children's favorite and one that, in skillful hands, provides a great deal of enjoyment. At first sight, the sphere looks like a large tennis ball with a cylindrical channel bored right through its center. However, when the string inserted in this channel is held at each end, the sphere will not drop suddenly down the string as it might be expected to do, but very gradually slides down, even stopping at times, to continue downward only when it is permitted.

Our illustration shows how this effect is produced: apart from the central channel, the sphere also has another, curved, one, whose two ends join the central channel very close to its ends. Those who know the secret will not, therefore, thread the string through the straight channel but through the curved one, and when the string emerges at the other end no one will be the wiser. The magician, for his part, need only tauten the string to a lesser or greater extent to delay the sphere's progress, or, indeed, to bring it to a complete halt, by increased friction.

On the left of our illustration we see the bewitched sphere floating in this way between the magician's hands, which are positioned one above the other.

If a wooden figure such as a clown, a juggler or an imp, was drilled through in the same way as the sphere, and substituted for it, the performance would be even more surprising.

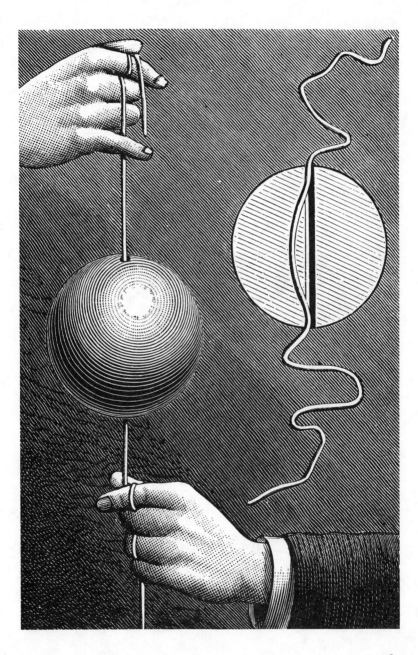

THE FRISKY MOUSE

ITINERANT peddlers often sell small mice which will run across the back of your hand as if they were real animals, tamed and trained. To stop them falling off his hand, the peddler will usually, before the "mouse" has reached the edge, place the edge of his other hand next to the first (see below), so that the small rodent moves from one hand to the next, and so on ad infinitum.

This fascinating and delightful game is based on an optical illusion. Let us look at it more closely.

The mouse, made of wood (or papier-mâché), is flat underneath and fitted with a little hook on top, around which the loop of a thread just over a foot in length has been wound, the free end being tied to a coat button. The mouse is placed on the back of the hand, which is slowly drawn downward so that it

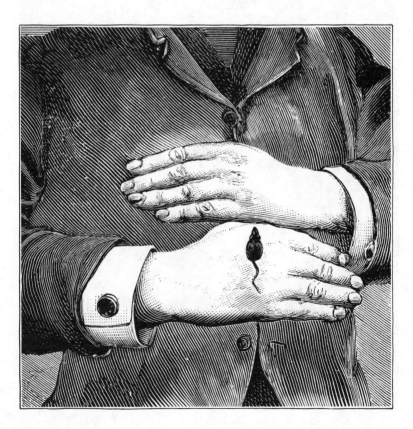

appears as if the little beast were moving across the hand in an upward direction. As soon as it reaches the edge of the hand and is about to fall off, the other hand is quickly placed over the first, whereupon the process can be continued.

This trick can be performed by anyone, since it requires no special equipment and involves very little expense.

The body of the mouse is carved from a piece of soft wood. The base is left flat, the two ears are fashioned out of pieces of cardboard, two small black beads are fitted for the eyes, the whiskers are made from a few bristles and the tail from a suitable piece of string; a small hook is fitted behind the head and the body is painted as true to nature as possible; finally a fine silk or other thread is attached to the mouse in the mannar described. And the rest follows, as we have said.

TRICKY DICE

THE dice trick described in what follows is based on the fact that two opposite sides of a die invariably add up to 7 points.

After two dice have been thrown on the table and the audience has been asked to count the points – let us say 5 (4 and 1) – the dice are picked up with thumb and index finger (Fig. 1 of our illustration). The opposite sides underneath now add up to 9, but the magician takes

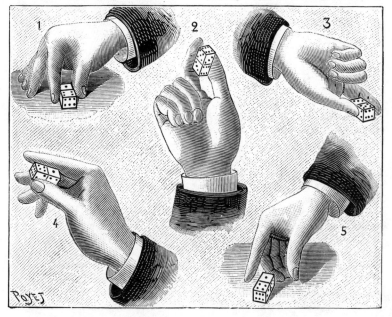

care not to demonstrate this fact. Instead, he tries to confuse the audience by giving the dice a quarter turn between his finger, so as to show 5 and 3 (Fig. 3), thus causing the spectators to conclude that they are in fact seeing the opposite faces to sides 4 and 1. After twisting his hand so as to bring the dice back to their original position, and again showing the spectators sides 4 and 1 (Fig. 4), the magician turns to his audience with the following words: "Ladies and gentlemen! You can see that the sides of these two dice add up to 5 and have satisfied yourselves that the underneath sides add up to 8. May I now ask you, Mr. So-and-so, to touch the underneath of the dice with this wand. Your touch will add one point to the sides. I shall now place the dice on the table and ask you to turn them over. How many points do you make it now?" " "Nine!" General astonishment.

"Now, ladies and gentlemen," the magician continues, "let us try it the other way around. Let us subtract instead of adding. Here we have two times six points [position 1]. I shall turn my hand over, to show you the reverse side [giving the dice a quarter turn once again]. What do they add up to now? As you see, the answer is nine. Now I would ask you, Mr. So-and-so, to touch the side, which you can see adds up to twelve [while moving his hand the magician has returned the dice to their original position), seven times with the magic wand. I shall now put the dice down on the table and ask you to turn them over. How many points are there?" "Two!"

The reader should be warned that there are some positions of the dice with which this trick cannot be performed. In such cases, he need merely drop the dice "accidentally" and start all over again.

The Clutch of Eggs in the Hat

This is a splendid trick. While making some appropriate remarks, the magician shows the audience a hat, an unfolded handkerchief and an egg, which he removes from a basket containing several eggs in full view of the audience. He then puts the basket to one side.

The selected egg is dropped into the hat, and the handkerchief quickly follows; when the handkerchief is pulled out again an egg appears as well. The handkerchief is put back in the hat, another egg is pulled out with it, and the process is continued as if the hat were full of

eggs. To prove that this is by no means the case, the hat can be turned upside down several times for, in fact, only one egg is involved in the trick, and those apparently removed from the hat are not really returned to the basket as the audience is made to believe.

The explanation is as follows: a thread about eighteen inches long is fastened with wax to a hollow, blown egg. It has a little hook at one end by which it is attached to the handkerchief, as shown in our illustration. The rest takes care of itself, for it is hardly

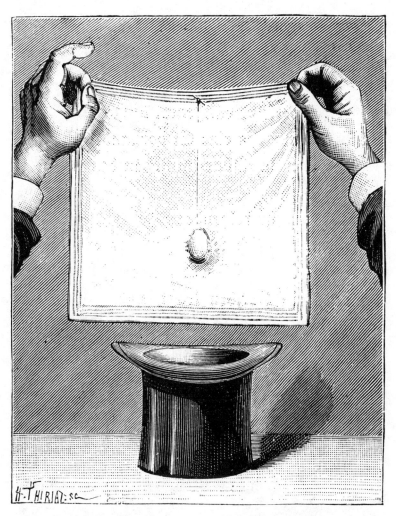

necessary to point out that only one egg is removed from the basket, that the empty side of the handkerchief is displayed, and that the two together are returned to the hat.

The egg may be used for several other tricks as well if the thread is long enough; when placed on the table it will, for example, move to and fro upon command, dance in time to music, balance on the end of a stick, and so on. A long horsehair may be used instead of a thread; it not only has the necessary fineness, but also considerably greater strength.

Flowers for a Friend

A LITTLE cylindrical container A (Fig. 1) is placed on the table together with a large cylinder C, which is closed at the top (Fig. 2). The magician holds the small container out for inspection to the audience, pours some soil into it, sprinkles on a few seeds, boldly claims that he will make the flowers grow, and places the cylinder over the container. After a suitable interval, he lifts up the cylinder to reveal a pretty nosegay in the small container.

What is the secret? Hidden inside the cylinder, made of cardboard or thin tin (Fig. 2 C), is yet another container, B, and it is the latter which is filled with the flowers. The cylinder, with B inside it, is pressed firmly over A, touched with the magic wand, and then raised, leaving A and B combined on the table. If the magician has a magic wand with a telescoping mechanism, he can (apparently) introduce the latter into the cylinder to persuade the public that there is nothing inside.

Simpler still is another way of conjuring up a bouquet, illustrated in Figs. 1 and 3. This time, an empty wineglass is placed on the table. The magician borrows a top hat and claims that, with its help, he will cause a bouquet to appear in the glass. While directing the attention

of the public to the glass, he inserts the middle finger of the hand holding the hat into the stemholder of the bunch of flowers kept in readiness beneath the edge of the table, conceals the flowers in the hat and brings the latter down over the wineglass, releasing his finger from the stemholder and placing the flowers, as required, in the glass.

THE MYSTERIOUS HAND

THE magician needs two chairs, across the backs of which he lays a large transparent sheet of glass. On it he places a hand in a cuff, fashioned of papier-mâché or soft wood. Hand, table and chairs are placed directly in front of the audience who can see that no mechanical devices are involved.

As soon as the hand is resting safely in the center of the glass plate, the magician steps back and begins to fire questions at this inanimate object. The hand immediately comes to life, and the fingers lift and rap several times on the glass, counting in this way, for instance, the points of one or several dice that are rolled onto the table, or indicating the ordinal position of various letters of the alphabet, or performing a host of similar feats. Needless to say the magician knows in advance how many points will show, for he uses

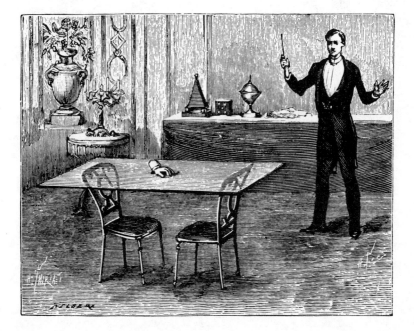

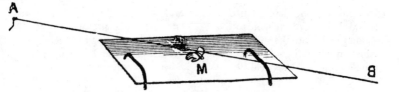

specially loaded dice which always end up in the same position, and naturally he has an assistant to whom he indicates, by carefully framed questions, what letter is on the table, or the age of a particular person, and so on.

The drawing above reveals the rest of the secret. A very thin thread of silk, rayon or the like is tied, say at *A*, at the rear of the stage. It runs right across the stage and ends in the wings at *B*. It is here that the assistant is hidden from public view and uses the string so skillfully that the spectators may well think the hand is alive and that the fingers are rapping by themselves.

THE PATIENT IMPALED

A DOCTOR sits on the stage. A patient comes in suffering from terrible stomach cramps, and is asked whether he has eaten or drunk too much, whether he has a cold, and so on. The case seems very compli-cated. "There is nothing for it," says the doctor, "we shall have to resort to a radical cure. Your stomach will have to be pierced and cleaned with a special tape." The doctor now seizes a six-foot blade with an eye through which a long ribbon has been threaded. The patient retreats in obvious panic. There is a great deal of arguing, with assurances from the doctor that the cure is perfectly harmless, and the patient putting up a fierce resistance until the very last moment, when the naked blade is thrust into his body. The doctor has sprung forward, plunging the blade in and pulling it and the ribbon right through the other side, and – miracle

of miracles! – the patient declares that the pain is completely gone.

The solution to this riddle is quite simply the device shown on the top right of our illustration. It is a hollow half-girdle made of metal, which the magician has fitted under the patient's vest. The two apertures at the ends of the girdle correspond to two button holes in the patient's special clothing, one in front and one at the back, so that the blade can go clean through. Since the blade is thin and very flexible, it fits easily into the curvature of the girdle. We said that the patient put up a fierce resistance when the doctor went for him. Needless to say, his real purpose was to fit the tip of the blade into the front button hole, whereupon the blade could follow its path around the body unimpeded. Since the blade becomes "shortened" by this process, it must be pulled "through" the body quickly.

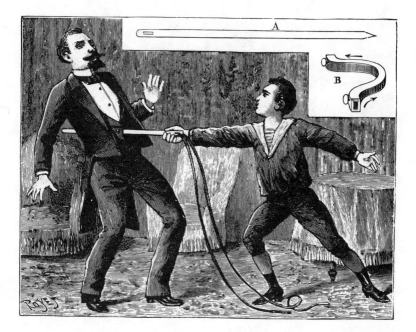

THE DISAPPEARING COIN

TAKE a large coin, mark it with some sign in full view of the audience, cover it with a cloth, and then ask one of the spectators to hold it over a glass of water (see the illustration on page 190, Fig. 1), asking him to drop the coin into the water as you pull the cloth away. He does as he is told, but the coin is not in the water! Claiming with great conviction that the coin must now be in the possession of one of the spectators, the magician steps up to the person in question and pulls the coin out of the astonished man's pocket, ear or where you will. He shows the mark on the coin to the audience, who are

sure that no exchange has taken place.

To perform this trick, we need a glass disk (Figs. 2 and 3 *A*) and a thin elastic band, to one end of which we have attached a small hook made from a bent pin, and to the other some wax. The elastic band is attached by its hook to the lining of the coat sleeve, and once the marked coin has been returned to the magician after inspection by the audience, it is pressed onto the sticking wax so that, when the free end of the band is released, the coin springs into the cuff. It is the glass disk, which must, of course, be the

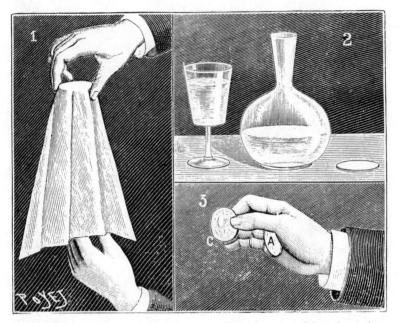

same size as the coin, that is covered with the cloth, and that the unsuspecting spectator actually drops into the glass of water. The transparent disk cannot be seen in the water and the coin thus appears to have vanished into thin air.

The discovery of the coin on the person of one of the spectators poses no problem. The magician simply pulls it unobtrusively out of his sleeve, removes the wax with a bent finger, and allows the coin to reappear where he chooses.

MAGIC WITH THE MAGIC EGG

WE began our ramble through the world of tricks, games and experiments beloved of our grandparents and great-grandparents with a balancing egg, and we shall now bring it to a close with a true piece of black magic.

In an eggcup stands an egg, that we cover with an egg cozy. We produce a silk scarf and draw it slowly into the hollow of both hands

(Fig. 1) until it disappears completely from sight. When we open our hands again, one is holding the egg, while the scarf has found its way into the eggcup (Fig. 3).

A good trick, and the reader is about to learn how it is done.

The egg in the cup is not real, and is, moreover, only half an egg. It is attached to the cup, which is open at the back, in such a way that it can be

flipped up and down. Inside this device a scarf has been hidden from the beginning, identical with the one we made disappear between our hands. Where did that one get to? It went into the egg which appeared in its place, and which was, in fact, made of metal, through an oval opening to be found in its back (Fig. 2 *F*). We are very careful not to show the opening, nor must we allow the egg to leave our hand.

We lift the egg cozy from the egg, and in its place the audience now sees the scarf (Fig. 3).

The solution is simple.

While we cover the cozy on the eggcup with our left hand, we quickly flip down the upper part of the cup with our right and the scarf makes its appearance. The cup must not be let out of our possession, either, lest some inquisitive person tries to examine it at closer quarters. It is best to perform another trick immediately after this one, so as to distract the attention of our audience.

Or we can bring down the curtain, lest someone accuse us of having fobbed him off with a rotten egg at the very end . . .

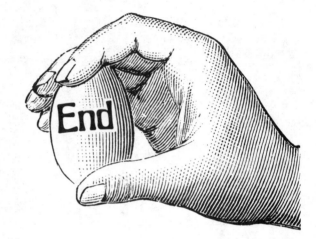